Burden of Proof, Presumption and Argumentation

The notion of burden of proof and its companion notion of presumption are central to argumentation studies. This book argues that we can learn a lot from how the courts have developed procedures over the years for allocating and reasoning with presumptions and burdens of proof, and from how artificial intelligence has built precise formal and computational systems to represent this kind of reasoning. The book provides a model of reasoning with burden of proof and presumption based on analyses of many clearly explained legal and nonlegal examples. The model is shown to fit cases of everyday conversational argumentation as well as argumentation in legal cases. Burden of proof determines (1) under what conditions an arguer is obliged to support a claim with an argument that backs it up and (2) how strong that argument needs to be to prove the claim in question.

Douglas Walton holds the Assumption Chair in Argumentation Studies and is Distinguished Research Fellow of the Centre for Research in Reasoning, Argumentation and Rhetoric at the University of Windsor, Canada. His most recent book is *Methods of Argumentation* (Cambridge University Press, 2013). Walton's work has been used to prepare better legal arguments and to help develop artificial intelligence. His books have been translated worldwide and he attracts students from many countries to study with him. A festschrift honoring his contributions, *Dialectics, Dialogue and Argumentation: An Examination of Douglas Walton's Theories of Reasoning and Argument*, edited by C. Reed and C. W. Tindale (2010), shows how his theories are increasingly finding applications in computer science.

Burden of Proof, Presumption and Argumentation

DOUGLAS WALTON

University of Windsor, Canada

CAMBRIDGE
UNIVERSITY PRESS

CAMBRIDGE
UNIVERSITY PRESS

University Printing House, Cambridge CB2 8BS, United Kingdom

One Liberty Plaza, 20th Floor, New York, NY 10006, USA

477 Williamstown Road, Port Melbourne, VIC 3207, Australia

314-321, 3rd Floor, Plot 3, Splendor Forum, Jasola District Centre, New Delhi - 110025, India

79 Anson Road, #06-04/06, Singapore 079906

Cambridge University Press is part of the University of Cambridge.

It furthers the University's mission by disseminating knowledge in the pursuit of education, learning and research at the highest international levels of excellence.

www.cambridge.org
Information on this title: www.cambridge.org/9781107678828

© Douglas Walton 2014

First published 2014

A catalogue record for this publication is available from the British Library

Library of Congress Cataloging in Publication data
Walton, Douglas N., author.
Burden of proof, presumption and argumentation / Douglas Walton,
University of Windsor, Canada.
pages   cm
Includes bibliographical references and index.
ISBN 978-1-107-04662-7 (hardback) – ISBN 978-1-107-67882-8 (paperback)
1. Burden of proof.   2. Proof theory.   I. Title.
K2263.W348   2014
347′.06–dc23   2014014939

ISBN 978-1-107-04662-7 Hardback
ISBN 978-1-107-67882-8 Paperback

For Karen, with love.

Contents

Acknowledgments

I would like to thank Springer-Verlag GmbH for permission to reprint three of my previously published papers. My paper "A Dialogical Theory of Presumption," originally published in *Artificial Intelligence and Law* (16 (2), 2008, 209–243), now appears in revised form as Chapter 3 of this book. A substantially revised version of my paper "Metadialogues for Resolving Burden of Proof Disputes," which appeared originally in *Argumentation* (21, 2007, 291–316), comprises parts of Chapter 6. My keynote lecture for the Sixth International Workshop on Argumentation in Multi-Agent Systems, originally published as "Burden of Proof in Deliberation Dialogs," in *Argumentation in Multi-Agent Systems*, ed. P. McBurney, I. Rahwan, S. Parsons and N. Maudet, Berlin, Springer, 2010, 1–22, has been reprinted in substantially revised form in this book, making up parts of Chapter 7. Finally, I would also like to thank the editors of *Informal Logic* for their permission to reprint materials from my paper, "Burden of Proof in a Modified Hamblin Dialogue System," which was originally published in *Informal Logic*, 31(4), 2011, 279–304.

I would like to thank Katie Atkinson and Tom Gordon for trenchant comments that were extremely helpful in making revisions. I would like to thank Henry Prakken for many detailed comments and corrections on technical matters in Chapter 2, and Erik Krabbe for helpful suggestions and comments on Chapter 5. I would also like to thank Joseph Laronge, Senior Assistant Attorney General of the State of Oregon, for drawing the ruling in the Lawson case to my attention, and for making many helpful comments and corrections.

The research for this book was supported by Social Sciences and Humanities Research Council of Canada Insight Grant 435–2012–0104: The Carneades Argumentation System. I am especially thankful to the members of the Centre for Research in Reasoning, Argumentation and Rhetoric (CRRAR) at the University of Windsor. They made many helpful comments when parts of some chapters were presented at CRRAR colloquia.

1

Introduction to Basic Concepts and Methods

A presumption is a device used in the law of evidence to enable a proposition to be taken into account as a piece of evidence in a case even though the argument supporting that proposition is not strong enough for it to meet a required burden of proof. From this definition of what presumption is, we can already see that presumption is linked to burden of proof in evidential reasoning in law. Burden of proof sets a standard for what is to be considered a proof in evidential reasoning in law. It is a device used to make it possible for a trial to arrive at a decision for one side or another in a contested case, even though all the facts of the case may not be known, and for various reasons may never be known. For example, in a criminal case, there may have been no witnesses to the crime, and the crime may have happened a long time ago. Most of the existing evidence may have been lost or destroyed. Therefore, evidential reasoning in law has to be able to move forward to a conclusion under conditions of uncertainty, lack of knowledge and even inconsistency. Typically, for example, in a trial there will be witnesses for one side, but there will also be conflicting testimony on the other side brought in by witnesses who say the opposite thing. What these conditions imply is that in a trial it is rarely if ever possible to prove or disprove the ultimate conclusion beyond all doubt. Hence, the device of having a burden of proof is necessary for the trial to reach a conclusion for one side or the other.

Presumption is not a new notion in legal reasoning. It was a device used in the ancient Jewish law code of the Talmud, and in ancient Roman law. A rough idea of how presumptions work is shown by citing some of the more common examples. According to the presumption of death, a person who has been unheard of for a fixed period of time, varying with the jurisdiction, five years typically in common law, may be presumed to be dead if there is no other explanation of his or her disappearance based on any evidence. Later in this book we will examine an example of another interesting kind of presumption called the presumption of mailing, which presumes that a properly addressed and stamped letter sent by the Postal Service was received by the person to whom it was addressed.

There are two purposes to this book. One is to explain and apply the latest methods of argumentation and artificial intelligence to help us understand how burdens of proof and presumptions work as devices of legal reasoning. The other is to use these resources to analyze burden of proof and presumption in everyday conversational argumentation. The task of describing and explaining how these models of burden of proof and presumption have been implemented in working software systems for analyzing and constructing legal arguments comprises a substantial part of the book. As argumentation has proved to be useful for artificial intelligence, this book will show how there is also a bounce-back effect enabling the benefits of the recent research in computer science to be applied to the problem of analyzing burden of proof and presumption. These two key concepts are fundamental to argumentation studies generally, and are known to be important for studying fallacies and other foundational problems that arise from the shifting of a burden of proof back and forth in a dialogue.

It is argued in this book that we can learn a lot from how the courts have developed procedures over the years for allocating and reasoning with burden of proof, and from how artificial intelligence models have built clear and precise logical models to represent this kind of reasoning. Indeed, the conclusion of the book, based on analyses of many legal and nonlegal examples, is that there is a general overarching structure of argumentation that fits cases of everyday conversational argumentation as well as argumentation in legal cases and that is based on an underlying common structure of burden of proof.

1. Problems and Objections

The concepts of burden of proof and presumption are fundamentally important in argumentation studies and indeed one could argue that they are the most fundamental concept in this area. These two concepts are so closely connected that it is impossible to study one without the other (Rescher, 2006). But procedural methods for dealing with issues of burden of proof and presumption in argumentation have been worked out and applied in most detail in the field of law. However, law itself is far from free of difficulties in being able to define and analyze this pair of concepts. According to Wigmore (1981, 285) the difficulties of every attempt to explain the concepts of burden of proof and presumption in law "arise not so much from the intrinsic complication or uncertainty of the situation as from the lamentable ambiguity of phrase and confusion of terminology under which our law has so long suffered." Kiralfy (1987, 94) wrote in a book written exclusively on the concept of burden of proof in law: "the phrase 'burden of proof' is notoriously difficult to define with any degree of precision, and when defined equally difficult to apply in a consistent way." In the standard work on evidence law *McCormick on Evidence*, Strong (1992, 449) wrote that presumption is the "slipperiest member of the family of legal terms," except for its first cousin, burden of proof.

There is considerable controversy concerning the extent to which legal methods for defining and determining burdens of proof can be applied to the study of problems of burden of proof arising in everyday conversational argumentation, and other context-like forensic debate. Hahn and Oaksford (2007) argued that the notion of burden of proof has been inappropriately extended into argumentation studies from its proper domain of application in law. They describe this extension as a "hasty transference" of legal concepts to less structured contexts of everyday conversational argumentation. Kauffeld (1998, 246) argued that the procedural formality of courtroom argumentation has been responsible for the lack of progress in investigating presumption and burden of proof in everyday conversational argumentation. Gaskins (1992, 3) even claimed that burden of proof works in law as a shadowy device used by skillful advocates in legal battles to direct manipulative arguments from ignorance against each other. On his view, public argumentation is deteriorating badly through the use of shadowy devices of burden shifting and arguments from ignorance. These objections are stated more fully in Chapter 1, but it won't be until the last chapter of the book that we can fully respond to them.

Hahn and Oaksford (2007) have argued that the notion of burden of proof has been extended inappropriately into argumentation studies from its proper domain of application in law. They call this extension a "hasty transference" of legal concepts to other kinds of argumentation, citing Gaskins (1992) and Kauffeld (1998) as supporting their view (2007, 40). On the account given by Hahn and Oaksford, Whately was the culprit who first affected the transference from law through the introduction of the notion of burden of proof in his writings on rhetoric. They also cite confusions and difficulties in the way the notion of burden of proof operates in law, citing the historical analysis of Gaskins (1992) to show how the U.S. Supreme Court of the Warren era used creative shifting of burden of proof as a vehicle for progressive social change (42). The two fundamental premises of Hahn and Oaksford's analysis are the propositions that burden of proof is only important where action is concerned, and that legal argumentation is about action. On their view (48), legal argumentation is characterized by a need for termination that arises from its inherent link to action. On their view, questions of evidence in law are subsidiary to decisions about actions. As well, on their view, "termination does not seem essential to argumentative dialogue in general"(48). On these considerations, they draw the conclusion (49) that there is no need for burden of proof in a critical discussion because it is not a type of dialogue with an inherent link to action.

As an example to support their case (2007, 43), they cite the decision that many countries have had to face when deciding whether or not to sign up for the Kyoto agreement. The majority of papers in leading scientific journals have accepted the claim that global warming is real, even though debate on the topic continues. However, they write (43), "the possible consequences of global warming are so potentially devastating that one might not want to wait until one was entirely certain before taking action." Accordingly, the procedure governments use is to set a threshold for action so that they can arrive at a decision when they are convinced enough to act. This example provides a paradigm case of the use of burden of proof as a device for rational decision making, leading to a course of action even under conditions of uncertainty. The problem posed by this kind of case is how burden of proof works in cases of deliberation where there is a need to take action and a choice has to be made on the evidence available. Is this different from the kind of case where the aim of a discussion is to find the truth by evaluating the evidence on both sides of a contested issue? This problem will be taken up in Chapter 7.

In the late 1960s and early 1970s, a new style of theorizing about evidential reasoning, called the New Evidence Scholarship, emerged in American law schools (Tillers, 1989, 1226). Some leading characteristic features of the New Evidence Scholarship can be summarized as follows. It focused more on logic as well as on law, it focused on the notion of proof in a way tying it to logical reasoning and it emphasized logical rigor as opposed to rhetoric. This scholarship also struggled with fundamental problems of epistemology, taking the approach that knowledge should be based on evidence rather than on justified true belief. Another characteristic was that the new scholarship employed technical tools from mathematics and formal logic, tools that were later also developed by artificial intelligence in computer science.

How research technical tools were used to formulate the outlines of a new theory of evidential reasoning and provide an approach that led to these later developments in artificial intelligence can be best appreciated by reading David Shum's book *The Evidential Foundations of Probabilistic Reasoning* (1994). His work defined the agenda of an important part of the new evidence scholarship (Tillers, 1989, 1226). Schum's work supported the view already widely accepted in law that the traditional approach to probability based on Bayesian rules do not take into account important features of the kind of reasoning used in realistic legal argumentation about evidence. Schum advocated and applied argument diagramming methods, of the kind that trace back to the use of diagrams to represent the mass of evidence on both sides in a legal case at trial called Wigmore charts (Wigmore, 1931). Wigmore's thesis was that there is an independent science of reasoning about evidence he called the "Science of Proof" that underlies the legal reasoning based on legal rules and procedures that we are familiar with. This thesis, when Wigmore first stated it, although it seemed generally like it should be true as an ideal, did not seem compelling as a program of research that could be realistically carried very far, because the science of reasoning was, at the time, confined to deductive logic and to the inductive kind of reasoning used in probability and statistics.

Instead of taking the standard Bayesian approach to probabilistic reasoning based on the study of games of chance in the Enlightenment period by scientists and academicians, notably Pascal, Schum took a different approach now called Baconian probability (Cohen, 1977). Baconian probability ties in closely with the new epistemological view of the New Evidence Scholarship approach. This epistemology defines knowledge as a defeasible concept that leads toward or away from the truth of the hypothesis being inquired into depending on the evidence supporting hypothesis and the evidence against hypothesis. On this approach, epistemology is closely tied to a cognitive model that evaluates a claim by considering both the arguments for it and the arguments against it, and by considering how the arguments for it interact with the arguments against it (in an orderly procedure). In the case of legal evidential reasoning, such a procedure might be, for example, that of a criminal or civil trial in the common law system.

The distinction between Pascalian and Baconian probability (Cohen, 1977; 1979; 1980) has become a matter of some importance for the study of legal argumentation in evidential reasoning. Each of these approaches to probability has a different kind of logic. For example, because evaluating argumentation in the approach of the New Evidence Scholarship requires examining and weighing both the proarguments and counterarguments, we are working in a system that needs to work with a knowledge base representing the mass of evidence in a legal case, allowing for inconsistency and incompleteness. This assumption has highly significant implications concerning how we should treat negation as well as negative evidence. In a Pascalian system, the probability value of a negated proposition is always calculated as unity minus the probability value of the original proposition. This probability rule will no longer work in any system of evidential reasoning based on Baconian probability. Similarly, when we put two pieces of evidence together as a pair of propositions joined by the conjunction operator, in Pascalian probability we basically multiply the probability values of the two simple propositions. This too will no longer work in evidential reasoning in law, for example where DNA evidence is used to corroborate or attack witness testimony evidence in a trial setting (Stein, 2005).

The Pascalian model is applicable to some instances of evidential legal reasoning, for example it is used in analyzing forensic DNA evidence. In recent times, however, there is a growing body of research in artificial intelligence and law that has gone beyond deductive and inductive logic (of the Pascalian sort) to use argumentation methods from informal logic that can be applied to defeasible reasoning under conditions of uncertainty, lack of evidence and conflicts of opinion where there are rules that apply but admit of exceptions. The methods are based on forms of argument that are subject to critical questioning and that only lead to conclusions that are tentatively acceptable subject to new evidence that may enter a case, and that sometimes fail. Such argumentation needs to be evaluated on a balance of considerations taking the pro- and contra reasons into account. We will see many examples of this throughout the book.

As these new methods were used in artificial intelligence tools and systems for evidential legal reasoning, Wigmore's thesis came to seem much more plausible and attractive. Legal reasoning was turning out to be a very good fit with argumentation methods because progress was being made in seeing how there is a common structure of reasoning or science of proof underlying both legal reasoning and everyday conversational argumentation. Hence, now is the time to see if we can go the other way and apply some of the lessons learned in artificial intelligence and legal reasoning to move research forward on some of the main concepts and problems of argumentation theory, both as it applies to law and other contexts of argumentation as well. One of the most significant concepts in this category is the notion of burden of proof, and with it the closely related notion of presumption. The central focus of this book in on the concept of burden of proof, but because the notion of presumption is so closely related to it (and indeed often confused with it), this concept has to come into detailed consideration as well.

2. Arguments from Ignorance

Gaskins (1992), in a broad social commentary that covers styles of legal reasoning as well as argumentation in everyday conversational contexts, has marshaled evidence that is supposed to show that the argument from ignorance has become "an inescapable feature of contemporary discourse" (3). He sees the argument from ignorance as forming the tacit structure of an increasingly common style of public argument: "I am right, because you cannot prove that I am wrong" (2). He has observed that this form of argument is found "in great abundance in public argument, in philosophical speculation, and throughout academic discussion" (2). According to his diagnosis, we live in a pluralistic age where we are increasingly insecure about resting arguments on fundamental principles, disciplinary foundations or a political notion of the common good (3). The consequence of our situation, according to Gaskins (3), is that there has been a polarizing tendency in public debate where each side deploys the argumentation strategy of attempting to impose the burden of ignorance on its opponent. According to his social commentary on the current state of affairs, the use of this strategy of arguing from ignorance hardens and exaggerates the difference between advocates on opposed sides of an issue. Each side declares, "I win, because you have not produced sufficient evidence to prove your point." On this view, the argument from ignorance, in these postmodern pluralistic times, has become a bad boy among argumentation styles that is running amok and distorting all our practices of public discussion, as well as scientific and legal argumentation. This claim appears to be that we are using argument from ignorance in these contexts much more than we used to, and this practice has had highly negative effects on these areas where argumentation is used.

To take one of his more dramatic examples, Gaskins (1992, 147) cites the case of the disastrous Challenger space mission in 1986, citing the view of a commentator who argued that the use of the argument from ignorance by NASA administrators was a main factor in the decision to go ahead with the launch. According to this description of the case, the basic philosophy of the manned space program had been associated with the principle, "Prove to me we're ready to fly." But in this instance, Gaskins argues, the logic of the situation was switched around by an argument from ignorance to the principle: "Prove to me we are not able to fly" (1992, 147). In effect, Gaskins is attributing the Challenger disaster to a tacit shift in the burden of proof effected through the use of the argument from ignorance.

Gaskins' claims that the argument from ignorance is powerful, dangerous and used commonly in both scientific and legal argumentation, as well as argumentation in everyday conversational discourse, have been abundantly confirmed by the study of many examples in (Walton, 1996). However, what has also been shown by this study of many examples of both reasonable and fallacious arguments from ignorance is that it is so extremely common in everyday conversational argumentation that most of us are unaware that we are using it so often. For a long time it seemed to be an exotic form of argument to those few people who studied logical fallacies, and it was assumed traditionally that it represented a fallacious form of argument. However, once its logical structure was revealed as having a characteristic argumentation scheme, it became possible to see that we are using it all the time to draw conclusions in cases where we have to reason from incomplete databases. But what these findings also reveal is that the claim that this form of argument is somehow especially characteristic of our argumentation in a pluralistic age where we are increasingly insecure about resting arguments on fundamental principles, disciplinary foundations or a political notion of the common good, is not very plausible, and would be impossible to prove. After all, if the argumentation scheme for argument from ignorance has been embedded in so many of the common arguments that we have always used since the earliest times when such arguments have been recorded, how can we prove, comparatively speaking, that its usage has spiked in these postmodern times? Once we begin to realize how common this form of argument is in all our reasoning, the hypothesis that it was not used as much before, but has now greatly increased in contemporary discourse, is open to question. It is an interesting idea for social commentary and speculation that the wide use of this form of argument, and the damage that strategically tricky uses of it can cause, has peaked in our pluralistic age. But how can we prove this idea as a hypothesis about changing styles of argumentation? Perhaps there is some clever way we could design an experiment to attempt to prove or disprove this hypothesis, but the hypothesis itself does seem highly dubious if argument from ignorance is as commonly used as the basis of our everyday reasoning, as well as scientific and legal reasoning, as the evidence so far suggests.

Gaskins links the argument from ignorance to the way burden of proof is used as a device in law. He characterizes burden of proof as "the law's response to ignorance, a decision rule for drawing inferences from lack of knowledge"(1992, 4). He describes the notion of burden of proof as vague and shadowy, operating in the background of legal procedure. He writes that in this respect, it is comparable to the default settings in computer programs. He describes it in negative terms as being often viewed by lawyers as a device for giving stage directions by determining procedural moves in legal argumentation, such as which party to a legal dispute has the obligation to speak first, and when such a party can step forward with evidence. This description of burden of proof makes it sound like a shadowy tool of legal argumentation that operates in the background and is wielded by lawyers and judges as a way of manipulating argumentation. He even writes that the wider influence of the notion of burden of proof on litigation "has been curiously ignored by legal commentators." This claim seems somewhat dubious, because there is a very large literature in evidence law on burden of proof, as well as a large literature on the related notion of presumption, and it is very well understood by legal scholars that the notion of burden of proof is fundamentally important, not only in evidence law, but in all legal argumentation generally. According to Gaskins (4) however, many legal standards are "notoriously vague" when applied to complex cases, and legal standards in such cases do not tell us where the burden of proof rests. He even goes so far to suggest that legal presumptions have been manipulated in order to orient the process of legal argumentation to favor judicial activism. As evidence of this claim he cites a number of·Supreme Court cases.

Despite what Gaskins says, when burden of proof and presumption are linked together, they function as evidential devices that are useful and even necessary when dealing with defeasible arguments that need to be used under conditions of uncertainty and lack of knowledge. Generally speaking, the burden of proof tells you how strong an argument needs to be in order to be successful. It represents a description of a task such that if you fail to carry out this task, your argument will fail. Burden of proof rests on the prior notion that there can be different standards of proof appropriate for different contexts of argumentation. This means that burden of proof might be discharged, making argument successful as a proof of its conclusion, even though the proof is not conclusive according to the requirements of some higher standard. To be realistic, we often have to make decisions based on evidence that cannot remove all doubt.

Burden of proof did not seem to be an important concept in mainstream philosophy in the past because it was generally assumed that in order for an argument to be successful it has to be a conclusive argument, in some sense meaning that it proves its conclusion beyond doubt. Perelman and Olbrecht-Tyteca (1971) and Toulmin (1964) showed that there was strong tradition tracing back to Descartes especially that favored certain knowledge based on conclusive proof that leaves no room for doubt. This was generally taken to mean that the argument had to be deductively valid and have premises that are known beyond doubt to be true. Reasoning based on probability, broadly of the statistical kind, was reluctantly allowed, but defeasible reasoning of the kind that only offered plausibility of a conclusion was seen as too subjective to be admitted as justification for rational acceptance. The impracticality of this view of the matter has long been implicitly recognized in law, where burden of proof is one of the most important factors in aiding courts to use reasoned argumentation to arrive at a conclusion. In typical cases of reasoning based on legal evidence, there is inconsistency and uncertainty in the evidence on both sides of a disputed issue, making a conclusive proof for one side an unrealistic requirement. For those of us seeking to grasp the structure of rational argumentation in a more realistic and practical way than the traditional methods of logic in philosophy made possible, there are many important clues to be found through the practical experiences of the courts, on how to develop and work with the notions of burden of proof and presumption. Unfortunately however, law itself has not found these notions entirely unproblematic to work with, and so there is much work to do to build some clear, consistent and coherent model of how burden of proof and presumption should be defined and should work in argumentation.

3. Three Examples of Burden of Proof Problems

In this section, three examples are presented that could be called classic cases of a problematic shifting of burden of proof from one side of a dialogue to the other. The first one took place in a parliamentary debate. The other two are both legal examples that went to trial, where the issue turned on burden of proof. In the political case, there was no resolution of the issue of which side should have the burden of proof, and the argument about the original issue of the debate simply carried on. In both legal cases, the court made a ruling on the issue of which side had the burden of proof. All three cases are instructive, but in different ways. Each brings out different aspects of how problems about burden of proof arise, and how they are resolved (or not) in different contexts of argument use.

The first example is part of a debate from the Canadian House of Commons that took place on September 30, 1985, described in Walton (1996, 118–120). The debate arose from concerns that an embargo on the export of Canadian uranium for nonpeaceful purposes was not being respected. It had recently been reported in the media that Canadian uranium was being used in American nuclear weapons. The question directed to the government representative was: "Can the minister give us the reasons why he is absolutely certain that depleted uranium is not being used for peaceful purposes?" The government representative responded as follows: "I have informed myself on the principle of fungibility and other arcane matters that are involved in this question. I have learned that there is, in the treaty, a requirement for administrative arrangements to be put into place that deal with the residue as well as with the original uranium. I have learned that those administrative arrangements are in fact in place. I am satisfied, on the basis of the information I know I have available, that the treaty is being respected." An opposition member then asked the question: "What is your proof?" The government representative replied: "I am asked for proof. The proof is that I have looked for any weaknesses in the treaty and I have found none. If honorable members have any information that the treaty is not being respected, I ask them for the fourth time not to be so secretive. Come forward with your allegations so that we can find out whether they are true or false." At that point, another opposition member said, "Do a proper investigation."

The sequence of argumentation in this case was classified in (Walton, 1996, 119) as fitting the argumentation scheme for the argument from ignorance, or *argumentum ad ignorantiam*, as it is traditionally called in logic. This form of argument, traditionally thought to be a fallacy, is often associated with shifts in a burden of proof (Walton, 1996, 58). The manual of rules for Canadian parliamentary debate (*Hansard*) does not define burden of proof. Procedural disputes, like those about burden of proof, are presumed to be resolved by the speaker of the House. In this case, the government representative began by replying that he investigated the matter, and was satisfied, based on his investigation, that the treaty was being respected. However, the opposition questioner, not satisfied with this standard of proof, asked him to give reasons why he is "absolutely certain" that the uranium is not being used for military purposes. This remark suggests an extremely high standard of proof, one which the government representative would be in no position to satisfy. The best the government representative could be expected to do would be to monitor violations, and be able to cite any that had been drawn to his attention, given the investigative resources at his disposal. At this point, the dialogue degenerates into an attempt by both sides to shift the burden of proof to the other side in a quarrelsome manner.

Another opposition critic even says, "Do a proper investigation." The problem is one of which side should have the burden of proof.

The second example is not from a trial or other legal setting. It is an ethical case, but is closely related to law. Wigmore (1981, 285) has a clear and simple example of the kind of argument we are all familiar with in everyday conversation argumentation in which burden of proof plays an important role. Wigmore thought that burden of proof in law operates in ways similar to the way it works in everyday conversational argumentation, but that there is also an important difference. Before discussing this difference, let us examine an example of a case of the operation of burden of proof in everyday conversation argumentation. It is interesting to note that the example is a three party dialogue. The two opposed parties, *A* and *B*, are at issue on any subject of controversy, not necessarily a legal one, and *M* is a third-party audience, or trier, who is to decide the issue between *A* and *B*. It seems to be a persuasion dialogue, for Wigmore tells us (285) that the desire of *A* and *B* "is to persuade *M* as to their contention."

Suppose that *A* has property in which he would like to have *M* invest money and that *B* is opposed to having *M* invest money; *M* will invest in *A*'s property if he can learn that it is a profitable object and not otherwise. Here it is seen that the advantage is with *B* and the disadvantage with *A*; for unless *A* succeeds in persuading *M* up to the point of action, *A* will fail and *B* will remain victorious; the 'burden of proof,' or in other words the *risk of nonpersuasion*, is upon *A*.

This example is used by Wigmore to show that the situation of the two parties is very different. The risk of failure is on *A*, because *M* will fail to carry out the action that *A* is trying to persuade him to carry out if *M* remains in doubt. Moreover, *M* will remain in doubt unless *A* brings forward some argument that will persuade him that investing in *A*'s property is a profitable object. In other words, *B* will win the dispute unless *A* does something. As Wigmore points out however (285), this does not mean that *B* is "absolutely safe" if he does nothing. For *B* cannot tell how strong an argument *A* needs in order to win. It may be that only a very weak argument might suffice. Therefore, to describe burden of proof in this example, Wigmore calls it the risk of nonpersuasion, describing it as "the risk of *M*'s nonaction because of doubt." The example shows that the burden of proof is this risk that falls on one side or the other in the dispute. In this example, it falls on *A*. This example is really a very good one to help us grasp in outline basically how burden of proof works in everyday conversational argumentation: "this is the situation common to all cases of attempted persuasion, whether in the market, home, or the forum" (Wigmore, 1981, 285). This case will be analyzed and discussed in Chapter 7, Section 4, and in Chapter 8, Section 2.

The normal default rule in law is the principle that the party who makes the claim has the burden of proof. This means, in a criminal case for example, that the party who has made the accusation that a crime has been committed has the burden of proving that claim. However, Anglo-American law (and other legal systems as well, such as the German and Dutch civil law systems) recognizes two different kinds of burden of proof, one called 'burden of persuasion' and the other called 'burden of production' (or 'burden of producing evidence,' or 'evidential burden'). According to *McCormick on Evidence* (Strong, 1992, 425), the term 'burden of proof' is ambiguous in law, covering both these two different notions. The burden of persuasion, allocated at the point in a trial where the judge instructs the trier (himself/herself or the jury) on what needs to be proved for the issue to be decided, is described (Strong, 1992, 425) as "the burden of persuading the trier of fact that the alleged fact is true." Unlike the burden of persuasion, the burden of producing evidence can shift back and forth from one party to another during the sequence of argumentation in a proceeding (Strong, 1992, 425).

Moving on to the third and fourth examples, both legal cases concerning burden of proof, we will start with the easier case and then move to the harder one. The easy case, ruled on by the U.S. Supreme Court in October 2005, began with a suit in a lower court (*Weast v. Schaffer*, 41 IDEL 176, 4th Cir. 2004). The parents of a disabled child, Brian Weast, sought reimbursement for private school tuition on the grounds that the program provided by their school district was inappropriate for his needs. Their argument was based on the Individuals with Disabilities Education Act, which requires school districts to create individual education programs for each disabled child. However, the act made no statement about the allocation of the burden of proof. The parents claimed that the district had the burden of proving that their program was appropriate, while the district held that the parents had the burden of proving that the district's program was inappropriate. When the case went to the U.S. Supreme Court, the starting point was the ordinary default rule that the plaintiff bears the burden of proof for the claim made. It was, however, acknowledged that school districts have a natural advantage in information and expertise, and that this imbalance might justify treating this case as an exception to the normal default rule. But it was argued that this exception did not apply in this case, because Congress had already obliged schools to safeguard the procedural rights of parents and share information with them. Parents have the right to review all records on their child possessed by the school, and the right to an independent educational evaluation of their child by an expert. The U.S. Supreme Court concluded that the burden of proof was properly placed on the parents.

This case was an easy one, because the parents' main argument was that putting the burden of proof on school districts will help to ensure that children receive a free special education for each disabled child. The U.S. Supreme Court concluded that this argument did not provide sufficient grounds for departing from the default rule on burden of proof. The U.S. Supreme Court also concluded that the exception on grounds of imbalance of expertise did not apply to this case. Therefore, the normal default rule automatically applied, and the conclusion drawn by the U.S. Supreme Court was that, in accord with this rule, the burden of proof is on the side of the parents. The next example is a harder case.

Prakken, Reed and Walton (2005) presented a hard case of a Dutch Supreme Court trial about the labor dispute in September 1980 in which a shift in the burden of proof in the case became the subject of a secondary dispute that threatened to deadlock the case. In this case, the band Los Gatos was hired to work for a cruise ship of the Holland America line. While the ship was waiting for repairs in harbor without passengers, the manager told the band to perform for the crew. The band refused, and the manager fired them. In Dutch law, this kind of dismissal is valid only if there was a pressing ground for it, for example, if the employee persistently refuses to obey reasonable orders. Los Gatos sued the Holland America line arguing that this pressing ground did not apply in their case because the Holland America managers had not wanted to listen to the reason why they refused to play. Neither party contended this assertion. Instead, the subject of the dispute was how much had to be proven by Los Gatos to adequately support their claim that their refusal to play was not a pressing ground for the firing. The issue was whether Los Gatos had to prove that they had a good reason to refuse to play, or whether Holland America had to prove that they did not have a good reason to refuse to play. In other words, it was a classic case of a dispute about burden of proof. The decision of the Dutch Supreme Court was that Holland America had the burden of proof. The reason given was that the Holland America managers had made it impossible for Los Gatos to explain their reasons for not wanting to play. This case is analyzed in detail in Chapter 6.

4. Survey of Theories of Presumption and Burden of Proof in Argumentation

There are two questions on burden of proof that the philosophical community disagrees about, and two schools of thought about how to answer them (Rescorla, 2009a, 86). If I assert a proposition and an interlocutor challenges me to defend it, who has the burden of proof? Am I always responsible for defending my assertion, or does the burden in some instances lie on the challenger? The dialectical foundationalists claim that some propositions do not require defense merely because someone challenges them. The dialectical egalitarians deny the existence of such propositions. In other words, they claim that if I assert a proposition, and my interlocutor challenges me to defend it, I am always responsible for defending it. According to Aristotle, for example, there are basic propositions that can be used for proving other propositions, but that cannot be proved themselves (Barnes, 1990, 120–123). Rescorla (2009a, 86) claims that this dispute recurs through the whole history of philosophy, and cites many leading philosophers, from Aristotle onward, who have defended the one school of thought or the other.

According to Rescorla (2009b, 87–88) the following rule of reasoned discourse, which he calls the defense norm, is widely, but not universally, held by philosophers: when challenged to defend an asserted proposition, one must either defend it or else retract it. Some philosophers, for example Brandom (1994, 177), claim that there are exceptions to the rule, like the assertions "There have been black dogs" and "I have ten fingers." The claim is not that such assertions are immune to doubt, but that when they are questioned, the question stands in need of justification. This claim itself is interesting, as it may suggest that sometimes questions need justifying and therefore there may sometimes be a burden of proof, or something like it, attached to the asking of a question. This would be a burden of questioning (a subject taken up in Chapter 5). These philosophers hold a qualified version of the defense norm Rescorla (2009b 88) calls the default challenge norm: when faced with a legitimate challenge to defend an asserted proposition, one must either defend it or else retract it. Many variants are possible on what constitutes a legitimate challenge (Rescorla, 2009b 89). The issues raised by these approaches take us to the question of how different norms regarding what constitutes proof, and different procedural rules for one can or must respond to an argument, should be set for different contextual settings where argumentation takes place. This quite general question will be addressed in the later sections of this chapter.

A survey of the most influential theories of presumption in argumentation theory has been presented by Godden and Walton (2007), beginning with the account given in Whately's *Elements of Rhetoric* (1846). Whately adopted the conservative position that there is a presumption in favor of prevailing opinions in existing institutions, such as the Church (1846, 114). The reasons why he adopted this conservative attitude may not be entirely clear, but his account of the connection between burden of proof and presumption is clear. According to his account, the burden of proof is initially placed on one side or the other at the outset of an argument. This initial placement has an effect on subsequent argumentation. The party who bears this burden has the responsibility of providing reasons in support of his position, and must give up that position if the reasons offered are insufficient or unsatisfactory. However, the raising of a presumption can relieve this burden and shift it from one side to the other.

Whately's account has often been criticized, and not only on the grounds that his conservative position seems to be a kind of special pleading in favor of religion (Whately was an archbishop of the Anglican Church). Critics like Kauffeld (2003) have argued that he basically does not provide clear criteria for the identification and justification of presumptive inferences, and that his analysis does not give a proper account of the foundation of presumptions because it retreats into notions of common sense and commonly accepted views. However, two features of Whately's account are noteworthy (Godden and Walton, 2007, 37). One is that he treats presumptions as subject to rebuttal, while the other is that, on his theory, presumption is closely tied to arguments from authority and expertise. Whately was often credited with basing his notion of presumption on principles of legal reasoning, but it has also been claimed that his theory is primarily psychological rather than legal in nature.

Alfred Sidgwick, a lawyer who wrote a well-known book on fallacies (1884), amplified Whately's view by writing (1884, 159) that "where a belief is in harmony with prevailing opinion, the assertor is not bound to produce evidence," but " whoever doubts the assertion is bound to show cause why it should *not* be believed" (Sidgwick's italics). However, Sidgwick was aware of the limitations of this view, and even remarked that Whately's presumption in favor of existing beliefs might amount to nothing more than an *argumentum ad populum*, a type of argument often held to be fallacious in logic. It might also be added that Sidgwick's account of presumption might amount to nothing more than an *argumentum ad ignorantiam*, an argument from ignorance or lack of evidence, another type of argument that has often been held to be fallacious in logic.

Kauffeld (2003) put forward a theory arguing that presumptions are justifiable on social grounds. According to his theory (2003, 140), to presume a proposition is to take it as acceptable on the basis that someone else has made a case for accepting it on the grounds that not accepting it will have the powerful negative social consequences of risking criticism, regret, reprobation, loss of esteem, or even punishment for failing to do so. A prominent feature of Kauffeld's theory is that it presents presumptions as similar to, or even coextensive with, social expectations (Godden and Walton, 2007, 322). On his theory, presumptions are grounded on rules of social conduct, which, if violated, bring a punitive effect on the violator. This approach could be questioned in its applicability to studying the logical aspects of presumption, as it seems to pay more attention to social and psychological factors than underlying inferential structures. However, as will be shown shortly, social expectations are important for understanding presumptions.

Ullman-Margalit (1983) recognized that there might be differences in the ways presumptions work in law and the ways they work in ordinary conversational reasoning. She suggested the research proposal of attempting to get a more refined and precise analysis of how presumptions work in ordinary reasoning by viewing them in light of the procedures already codified and widely studied in law. The outcome of her analysis was to define presumption in terms of the characteristic sequence of reasoning from premises to a conclusion. There are three parts to the form of inference defining the sequence (1983, 147). The first part is the presence of the presumption-raising fact in a particular case at issue. The second part is the presumption formula, which sanctions the passage from the presumed fact to a conclusion. The conclusion is that a proposition is presumed to be true on the basis of the first two parts of the inference structure. She is very careful to describe the status of the conclusion of this presumptive inference, writing (147) that the inference is not to a "presumed fact," but to a conclusion that "a certain fact is presumed."

Ullman-Margalit emphasized the practical nature of presumption and its connection with argumentation from lack of evidence. She described presumptions as guides useful for practical deliberation in cases where there is an absence of information or conflicting information that interferes with the formation of a rational judgment but where nevertheless, some determination must be made in order for an investigation better to proceed (152). She emphasized that presumptions are not always justified, and enunciated the principle that the strength of a presumption in a given case should be determined by the strength of its grounds on a case-by-case basis (157). She also emphasized the inherent defeasibility of presumptive rules, stating that such a rule contains a rebuttal clause specifying that it is subject to exceptions (149). All these characteristics turn out to be important in the new dialectical theory proposed in this book.

The dialectical theory of presumption put forward by Walton (1992) was meant to be applied to everyday conversational argumentation. It was not specifically addressed to how presumption works in legal argumentation. According to this theory, in conversational argumentation presumptions take the form of cooperative conversational devices that facilitate orderly collaboration in moving the resolution of a dispute forward even if not everything can be proved by the evidence available.[1] A context of dialogue involves two participants, a proponent and a respondent. The dialogue provides a context within which a sequence of reasoning can go forward with a proposition A as a useful assumption in the sequence. The principle of adopting a presumption in a conversational exchange has the form of a dialogue rule that appears to violate the usual requirement of burden of proof: even if there is no hard evidence showing that a proposition can be proved true, it can be presumed (tentatively) true, subject to later rejection if new evidence proves it false. On this theory, the key characteristic of presumption as a speech act in dialogue is that it reverses an existing burden of proof in a dialogue by switching the roles of the two participants. Normally, the burden of proof is on the proponent asserting a proposition, but when a presumption is activated, this burden of proof shifts to the respondent,

---

[1] Note that on this dialectical theory, presumptive reasoning has a negative logic, and is therefore closely linked to lack of evidence reasoning.

once the presumption has been accepted as a commitment in the dialogue. In this dialectical theory, the point where the presumption is first brought forward in a dialogue is called "move x," while the point where it may be rebutted is called "move y." This working of a presumption is regulated by the following key seven dialogue conditions, summarized from the fuller list in (Walton, 1992, 60–61).

C1. At some point x in the sequence of dialogue, A is brought forward by the proponent, either as a proposition the respondent is asked explicitly to accept for the sake of argument, or as a nonexplicit assumption that is part of the proponent's sequence of reasoning.
C2. The respondent has an opportunity at x to reject A.
C3. If the respondent fails to reject A at x, then A becomes a commitment of both parties during the subsequent sequence of dialogue.
C4. If, at some subsequent point y in the dialogue ($x < y$), any party wants to rebut A as a presumption, then that party can do so provided good reason for doing so can be given.
C5. Having accepted A at x, however, the respondent is obliged to let the presumption A stay in place during the dialogue for a time sufficient to allow the proponent to use it for his argumentation (unless a good reason for rebuttal under clause C4 can be given).
C6. Generally, at point x, the burden of showing that A has some practical value in a sequence of argumentation is on the proponent.
C7. Past point x in the dialogue, once A is in place as a working presumption (either explicitly or implicitly) the burden of proof falls to the respondent should he or she choose to rebut the presumption.

Applying this theory of presumption enables a dialogue to move forward by giving the argumentation a provisional basis for moving ahead, even in the absence of sufficient evidence to prove key premises. How such presumptions should be accepted or rejected in a given case is held to depend on the type of dialogue, the burden of proof set at the beginning of the dispute and factors in specific arguments like argumentation schemes. Walton's account contrasts with Ullman-Margalit's to some extent, as hers appears to be more inferential in nature while his appears to be more explicitly dialectical in nature.

Hansen (2003) proposed an inferential analysis of the structure of presumptive inference that is comparable to that of Ullman-Margalit in that a presumption is always taken to have three parts: a major premise that expresses a rule, minor premise that expresses an antecedent fact, and a conclusion stating a presumption drawn by combining the major and minor premises. However, instead of basing his account on legal reasoning, Hansen based it on Whately's theory that presumptions in ordinary reasoning are inferred from presumptive rules using this three part structure.

Rescher's theory brings the Ullman-Margalit and Walton theories together by making an integrated theory in which presumption has two components that fit together. The first is the dialectical component, meaning that presumption is defined in relation to formal structure of disputation of the Rescher type in which there are three parties. The second is the logical component, in which presumption is defined in relation to a certain characteristic type of logical inference. The latter rests on Rescher's defining principle for an appropriate cognitive presumption (2006, 33), which has the form of a general rule: "Any appropriate cognitive presumption either is or instantiates a general rule of procedure of the form that to maintain *P* whenever the condition *C* obtains unless and until the standard default proviso *D* (to the effect that countervailing evidence is at hand) obtains." *P* is the proposition representing the presumption.

Rescher (2006) at first appeared to be taking up Ullman-Margalit's program of research, when he characterized presumption by outlining the historical development of the use of the concept in law, stating that presumption has figured in legal reasoning since classical antiquity (2006, 1). However, his theory is much broader in its intended applications. It is by no means restricted to explaining how presumptions work in law, or even in everyday reasoning. He also investigates presumption in science and in economic and political decision making. He takes inquiry and deliberation into account, as well as persuasion dialogue. Rescher (1977) also appears to have been the first to develop a detailed account of presumption in an explicitly dialectical framework, drawing both on formal models of disputation and the legal origins of the notion of presumption in burden of proof (Godden and Walton, 2007, 324). Rescher wrote (1977, 25) that burden of proof is a legal concept that functions within an adversary proceeding where one side is trying to prove a charge while the other is trying to rebut it before a neutral trier of fact. An especially distinctive feature of his way of analyzing burden of proof using a formal dialogue model is that three parties are involved, a proponent and an opponent who put forward arguments and rebuttals, as well as a third party trier who sees that proper procedures are followed and decides the outcome of the disputation.

Rescher (1977, 27) drew a distinction between two different types of burden of proof. First there is the probative burden of proving an initiating assertion, stating that an advocate of a claim in a dialogue has the burden of supporting it with argument. Second there is "the evidential burden of further reply in the face of contrary considerations." He calls the second type of one of "coming forward with the evidence" (27). It appears to correspond to what is usually called the burden of producing evidence in law, or the burden of production. Thus, it would seem that Rescher's account roughly parallels the two main legal notions of burden of proof (Godden and Walton, 2007, 325). On Rescher's account, presumption is closely related to burden of proof, to rules and to argument from ignorance. The latter connection is particularly evident when Rescher (2006, 6) writes that a presumption is not something that "certain facts *give* us by way of substantiating evidentiation," but rather something that "we *take* through a lack of counterevidence" (Rescher's italics). It appears that he primarily refers to defeasible rules of the kind that are subject to exceptions,[2] and thus in cases where such rules are used to support arguments, it would be expected that in a dialogue, arguments and rebuttals would go back and forth from one side to the other. This is in fact the standard format in any formal model of dialogue modeling disputation, including Rescher's. Presumption is described in such a format as a device that "guides the balance of reasons" in the shifting of the burden of proof from one side to the other during a disputation. On this account, "a presumption indicates that in the absence of specific counterindications we are to accept how things as a rule are taken as standing" (1977, 30). Thus, if there is a general rule that when brought into play favors the argument of one side, a presumption is a device that uses the rule to shift the burden of coming forward with evidence against the other side.

Another feature of Rescher's theory worth noting here is that there are three especially significant kinds of grounds determining on which side a presumption lies in a dialogue. One such ground is negotiated agreement. A second, reminiscent of Whately, is the standing of an authoritative source (Rescher, 1977, 39). A third important one is plausibility, for presumption, we are told, generally favors the most plausible among a set of alternatives (38). Note that plausibility on Rescher's account often depends on how things can normally be expected to go in a familiar situation, in a way that is reminiscent of Kauffeld's theory.

5. Presumption and Burden of Proof in Legal Argumentation

It is often said that the burden of proof shifts back and forth from one side to the other during the sequence of argumentation in a trial as evidence is put forward by either side. It is often also said that it is the device of presumption that shifts the burden of proof from one party to the other during a trial. The notion of presumption also appears to be directly connected to what is called the presumption of innocence in the criminal trial. This device sets the burden of proof much higher on one side than the other. The prosecution has to prove that the defendant committed the crime he was alleged to have committed to the standard of beyond a reasonable doubt, whereas the defense has only has to raise enough doubt to show that the prosecution did not fulfill this burden. If the prosecution has only presented a sequence of argumentation that is too weak to meet the beyond

---

[2] Rescher (2006, 6) specifically states the idea of presumption is closely linked to the notion of defeasible reasoning (default position) in computer science.

reasonable doubt standard, theoretically the defense doesn't have to say anything at all in order to win the trial. The rationale for this asymmetrical way of setting the burden of proof in a criminal trial in law is that it is better to let many guilty parties go rather than to unfairly convict an innocent person for a crime he did not commit. Because of the nature of the evidence in criminal cases so often being incomplete so that a conclusion beyond all doubt is not possible, there will inevitably be some innocent defendants convicted. But the idea behind the asymmetrical setting of burden of proof is that this number should be minimized to the extent possible by the evidential rules and standards of proof set in a criminal trial.

In *McCormick on Evidence* (1992, 449) Strong wrote that presumption is the "slipperiest member of the family of legal terms," except for its first cousin, burden of proof. Encouragingly, however, several recent studies of burden of proof and presumption have appeared in artificial intelligence and law (Prakken, Reed and Walton, 2005; Prakken and Sartor, 2006; Gordon, Prakken and Walton, 2007; Prakken and Sartor, 2007) that offer formal models that can render these important but slippery and vague notions into precise tools useful for helping us to analyze and better understand the roles of presumption and burden of proof in legal reasoning.

The following example can be used to show how burden of proof can shift in a murder trial, but it is expressed in relation to how the crime of murder is defined in a specific set of rules for criminal law. Murder is defined as unlawful killing with malice aforethought in section 197 of the California Penal Code. Section 187 defines an exception for self- defense. In the example, there is sufficient evidence to prove the killing and malice elements of the crime are based on sufficient evidence so that the defense has accepted these premises. Next, the defense puts forward an argument for self-defense, by calling a witness who testified as that the victim attacked the defendant with a knife. But in the next sequence of argumentation in the example, the prosecution calls another witness who testifies that the defendant had enough time to run away.

How does this argumentation affect the burden of proof? To begin with, the prosecution has the burden of persuasion in a criminal case. But after the defendant has met his burden of production for self-defense, the proof standard for the self-defense statement is changed to a standard that reflects the prosecution's burden of persuasion because the standard is satisfied only if the best con argument has priority over the best pro argument. While the prosecution is the proponent of the main claim, namely the murder charge, the defense is the proponent of the exclusion by the self-defense rule. The defense is also the proponent of the claim that the defendant did act in self-defense, but due to the prosecution's burden of persuasion in a criminal case, it has the evidential burden of persuading the trier that the defendant did not act in self-defense.

A trial can be modeled as a dialogue between two opposed parties in which each side puts forward arguments to support its claim, and the effect of bringing forward an argument based on evidence possibly including a presumption is to shift an evidential burden from one side to the other in the dialogue. This dialogical approach to legal argumentation would handle this example by saying that presumption is a kind of move in a dialogue different from the move of making an assertion or claim based on evidence sufficient to support it. To presume that a proposition is true is to request the other party in a dialogue to accept it without having to give sufficient evidence to support it and fulfill the normal kind of burden of proof that would be required to back up an assertion.

An example (Prakken and Sartor, 2006) is a case where the plaintiff demands compensation on the ground that the defendant damaged his bicycle. The plaintiff has the burdens of production and persuasion that the bicycle was damaged and that he owned it. One way he can prove that he owns the bicycle is to prove that he possesses it. According to Dutch law in such a case, given possession, ownership of the bicycle can be presumed. The presumption in such a case can be expressed by the proposition that possession of an object can be taken as grounds for concluding that the person who possesses the object owns it. According to the Prakken-Sartor theory, this proposition has the form of a default rule, and generally speaking, any legal presumption can be cast in the form of such a default rule. The default rule is this proposition: normally if a person possesses something, it can be taken for granted that he owns it, subject to evidence to the contrary. It is held to be a default rule in the Prakken-Sartor theory in the same way the following proposition is: if Tweety is a bird, then normally, but subject to exceptions, Tweety flies. Such a proposition is a default rule in that it holds generally, but can fail or default in the case of an exception, for example in the case that Tweety is a penguin.

According to Prakken and Sartor (2006, 23–25), there are three types of burden of proof that need to be distinguished carefully in law, called burden of persuasion, evidential burden and tactical burden of proof. The burden of persuasion rests on a party in a trial, or comparable legal proceeding, and it requires that this party must prove a designated proposition by supporting it with grounds that are sufficient for endorsing it at the end of the trial. This proposition is called the ultimate *probandum* of the trial, the ultimate proposition to be proved. For example in Dutch law, to prove the case of alleged manslaughter, the prosecution needs to satisfy its burden of persuasion by proving that the defendant killed the victim with intent (23). Killing and intent are often called the elements of the ultimate *probandum*. To fulfill its burden of persuasion, the prosecution has to prove that the defendant not only killed the victim but did so with intent. This burden of proof does not change over the whole course of the trial, and it is fulfilled or not only in the final stage when the jury decides the outcome of the trial.

In contrast with the burden of persuasion, the evidential burden and the tactical burden are often said to shift back and forth during the course of the trial from one side to the other. In Dutch law (Prakken and Sartor, 2006, 24), the accused can only escape conviction by providing evidence of an exception to the rule that if killing and intent are proved, the defendant is guilty of manslaughter. One exception of this sort would be evidence that the killing was done in self-defense. Such evidence could be provided if the defendant could provide a witness who claims the victim threatened the accused with a knife. However, the defense does not have to prove self-defense, by a standard of proof that would be suitable to fulfill a burden of persuasion. All it must do is produce some evidence, enough evidence to raise the issue of self-defense, and it throws sufficient doubt on whether the judge should rule that there is no self-defense. This type of burden can be called the evidential burden, but it is also often called the burden of production, or the burden of producing evidence.

There is a third kind of burden of proof that Prakken and Sartor call the tactical burden of proof. Suppose the defense presents enough evidence to fulfill the evidential burden for a finding of self-defense and the prosecution attempts to rebut this argument by bringing forward a witness who declare'd that the defendant had enough time to run away. If the prosecution's argument is strong enough, it would have the effect of making the prosecution's ultimate *probandum* of manslaughter justified once again. This move puts a tactical burden of proof on the prosecution. They might discharge it, for example, by arguing that the witness put forward by the prosecution is a friend of the victim, and that this fact makes her an unreliable witness. Accordingly, a tactical burden of proof can shift from one side to the other, as each side brings forward a new argument. Prakken and Sartor argue (2006, 25) that in contrast, the burdens of production and persuasion are fixed and cannot shift from one party to the other. This claim is clearly true for the burden of persuasion, which remains on a party until the last stage of the trial. However, it seems less clear that the evidential burden is fixed in this way. The reason that Prakken and Sartor give to support their claim that the evidential burden is fixed is that this burden on an issue "is fulfilled as soon as the burdened party provides the required evidence on that issue and after that is no longer relevant." It should be remarked here that there appears to be considerable disagreement and even controversy on the question of whether the evidential burden shifts back and forth. Most legal commentators appear to assume that it does often shift back and forth from one side to the other in a trial, but some commentators, including Prakken and Sartor, have argued that it never does. These disagreements need to resolved.

6. Shifting of Burden Proof and Critical Questioning

The basic argumentation scheme for argument from expert opinion is presented below using the notation from (Walton, 1997, 210). The variable E stands for an expert, whether it is a human expert or expert system knowledge base. A is a propositional variable. We use the term "proposition" as being equivalent to the term "statement."

Major Premise: Source E is an expert in subject domain D containing proposition A.

Minor Premise: E asserts that proposition A (in domain D) is true (false).

Conclusion: A may plausibly be taken to be true (false).

This argumentation scheme for argument from expert opinion is best seen as a defeasible form of argument, as indicated by the phrasing in the conclusion. The philosophy behind this way of conceiving the scheme is based on two assumptions. The first one is that it is not justifiable to take the word of an expert as infallible, or to defer to an expert without questioning what she says. The second one is that because experts tend to be more right than laypersons in a given field, there can be good reasons in many instances to accept what an expert says. With this approach, accepting expert opinion in a cautious and qualified manner can sometimes be a reasonable way to move ahead under conditions of uncertainty and lack of knowledge. The qualification, however, is that the arguer moving ahead on the basis of expert opinion must be prepared to critically question the advice given by the expert if there are reasons to doubt it. That is why there is a set of critical questions attached to the argument from expert opinion.

The following six critical questions are the standard ones for the argumentation scheme for argument from expert opinion given in Walton (1997, 223).

1. Expertise Question: How credible is E as an expert source?
2. Field Question: Is E an expert in domain D that A is in?
3. Opinion Question: What did E assert that implies A?
4. Trustworthiness Question: Is E personally reliable as a source?
5. Consistency Question: Is A consistent with what other experts assert?
6. Backup Evidence Question: Is E's assertion based on evidence?

The first question concerns the depth of knowledge the expert may be assumed to have, depending on how much experience the expert has in the domain, and other factors that may be used to judge depth of expertise. The problem confronted by the second critical question is that there is often a halo effect, so that an expert in one domain may be taken as such a respected authority that what she says in another domain that is unrelated to the first one may be accepted more strongly than the evidence merits. The third question probes into the exact wording of what the expert said, and concerns matters like whether what the expert said was quoted or paraphrased. The fourth question concerns the trustworthiness of the expert. For example, an expert that is biased toward one side of an issue may be seen rightly as less trustworthy than an expert that has not exhibited any bias. The fifth concerns the evidence the expert's pronouncement is based on. If an expert is requested to provide evidence to back up her statement, she should be willing to provide that evidence, and if she fails to provide evidence when it is requested, that is a reason for having doubts about the acceptability of her statement.

The normal procedure in standard formal dialogue systems is to put the burden of proof on the proponent of any argument, while there is no burden on the respondent to put forward counterarguments, or to question the proponent's argument. However, such a one-sided approach seems limiting, because if the respondent merely accepts the proponent's argument, the dialogue as a whole might be a failure. For example, in the persuasion dialogue, if the respondent fails to probe deeply enough into the proponent's arguments and find critical flaws and failures in it, the proponent will receive inadequate guidance on how to correct these failures and thereby improve his arguments. The same one-sidedness arguably applies to legal argumentation in a criminal trial in the common law system, where the burden of persuasion lies on the prosecution (Walton, 2003). If the defense fails to question an argument, for example, an argument from expert opinion, the argument stands, even though it could later be questioned in the examination stage.

One might reply that there is no need for a burden of questioning either in everyday persuasion dialogue or in the more regulated setting of a legal trial. The reason is that if the respondent fails to question critically an argument put forward by the proponent, the penalty for this failure will simply be that the respondent will lose the dialogue. And hence, at least indirectly, this burden of loss will be enough encouragement to make the respondent critically question the proponent's arguments. Still, however, it has been considered whether a rule requiring the respondent to critically question the proponent's argument could be useful in some instances, or in some types of dialogue. Such a rule would require the respondent give a reason for doubting an argument, and for not accepting its conclusion, by asking appropriate critical questions or putting forward a counterargument. However, the normal kinds of persuasion dialogues presented so far in the literature have no burden of questioning.

Recent work in argumentation (van Laar and Krabbe, 2012) is only now beginning to examine whether there should be a burden of questioning on the respondent in a reasoned dialogue as well as a burden of proof on the proponent of an argument. They hypothesize that the competitiveness inherent in the persuasion type of dialogue needs to be mitigated by making the respondent responsible for putting forward counterarguments or critical questions, if they are available. The reason that they give is that the quality of the discussion can be enhanced if the participants behave in a more cooperative way by offering advice to the other party in the form of providing questions or counterarguments that enable the proponent to see the weaknesses in his argument and improve it as the discussion proceeds.

Despite the potential advantages of modifying formal dialogue systems to represent this possibility, in this book, the discussion of burden of proof has been restricted to dialogue systems in which there is no burden of questioning in addition to the burden of proof on the proponent to prove his claims. There are two reasons for this restriction in the scope of the book. One is that building systems with special rules for a burden of questioning makes for more complex systems with additional rules. However, because this book has enough to do to struggle with what are taken to be the traditionally more simple requirements on burden of proof, the decision was made not to include considerations of this sort. The second reason is that in the main formal model we will study, the Carneades Argumentation System, critical questions matching an argumentation scheme are treated by making them into additional premises of the argument called assumptions and exceptions. How this procedure enables us to deal with burden of proof regarding the asking of critical questions is explained in section 7.

When an argument from expert opinion is put forward, a burden of questioning is placed on the respondent to either accept the conclusion of the argument provisionally or raise critical questions about it. When such a critical question is asked by the respondent, the burden of proof is shifted back onto the proponent of the argument to answer the question appropriately. Otherwise, the argument from expert opinion defaults. The key problem in representing how this procedure of critical questioning should be uniformly modeled in an automated argumentation system is that there are two theories on how the burden of proof should shift back and forth. According to one theory, called the shifting initiative (SI) theory, merely asking the critical question is enough to defeat the argument until the critical question has been answered appropriately. According to another theory, called the backup evidence (BE) theory, merely asking the critical question is not enough, and in order to make the argument default, some evidence needs to be brought forward to support the critical question. The problem of how to distribute burden of proof in critical questioning of an argumentation scheme has been intensively studied. The outcome is that each scheme has to be examined on its own merits, and a decision made with respect to that scheme on the matter of classifying each critical question into the one category or the other with respect to shifting of the burden of proof. This problem was solved by the Carneades Argumentation System, and to see how the solution was implemented we have to give a brief outline to explain how this system works.

7. The Carneades Argumentation System

In the Carneades system, critical questions matching an argumentation scheme are modeled as premises of the scheme, some of them being additional to the premises explicitly given in the scheme. These premises are classified into three categories, called ordinary premises, assumptions and exceptions. The ordinary premises are the ones that are explicitly stated in the argumentation scheme. The assumptions are additional premises that , are assumed to hold, just like the ordinary premises, but if questioned by the asking of a critical question they automatically fail to hold unless the proponent of the argument gives some evidence to support the premise. In other words, assumptions fit the SI theory. They shift the initiative back on to the proponent as soon as the question is asked. Exceptions are also additional premises except that they are assumed not to hold unless evidence is given by the critical questioner to show that they do hold in the case at issue. This means that exceptions fit the BE theory. They do not defeat the argument unless the questioner gives backup evidence to support the question.

This system enabled a solution to the burden of proof problem with critical questions to be put into place by enabling the critical questions to be represented as different kinds of premises of an argumentation scheme represented on an argument diagram. Following this approach, all the critical questions matching the argumentation scheme for argument from expert opinion were classified as shown in Walton and Gordon (2005).

Ordinary Premise: E is an expert.
Ordinary Premise: E asserts that A.
Ordinary Premise: A is within F.
Assumption: It is assumed to be true that E is a knowledgeable expert.
Assumption: It is assumed to be true that what E says is based on evidence in field F.
Exception: E is not trustworthy.
Exception: What E asserts is not consistent with what other experts in field F say.
Conclusion: A is true.

The trustworthiness question raises doubt about the reliability of the expert as a source who can be trusted. But unless there is some evidence of unreliability, the proponent could simply answer "yes," and that would seem to be enough to answer the question appropriately. To make such a charge stick, the questioner should be held to supporting the allegation by producing evidence of bias or dishonesty. If the expert was shown to be biased or a liar, that would presumably be a defeater. It would be an *ad hominem* argument used to attack the original argument, and if strong, would defeat it. But even raising the question of trustworthiness represents quite a strong attack that makes a serious allegation demanding some kind of proof. For comparable reasons, the consistency critical question was classified as an exception. However, because it is generally assumed that experts are knowledgeable, and that what they say may be presumed to be based on evidence in their field or domain of expertise, these other two questions are classified as assumptions.

The recognition that argumentation skills can benefit greatly from the use of computational tools to support and teach argumentation has led to the development of argumentation software. One of these tools, called Carneades, is domain-independent, but primarily aimed at legal argumentation. Carneades is an Open Source software project that provides tools for supporting a variety of argumentation tasks including argument mapping and visualization, argument evaluation, applying proof standards respecting the distribution of the burden of proof and argument construction from rules and precedent cases. Carneades supports and uses argumentation schemes, including the one for argument from expert opinion.

The Carneades Argumentation System is based on mathematical structures and definitions (Gordon, 2010) and the system has been implemented as an open source software program that is freely available on the Internet (http://carneades.github.com/). It has a graphical user interface so the user can draw argument diagrams showing inferential relationships among premises and conclusions of arguments. The system is domain independent, meaning that even though it is especially designed to model legal argumentation, it can easily be used to draw argument diagrams representing arguments used in everyday conversational reasoning.

To reconstruct and evaluate an example of an argument by applying the Carneades argument mapping tool, the user has to go through a sequence of steps:

1. Make a list, called the key list, of all the propositions in the argumentation.
2. Identify the premises and conclusions of each argument using the propositions in the key list.
3. Connect the sequence of argumentation into a chain of arguments in which the conclusion of one argument can be a premise in another one.
4. Assign a proof standard, such as preponderance of the evidence, to each proposition.
5. If an argument fits a known argumentation scheme label the argument with the name of that scheme.

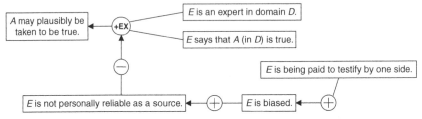

FIGURE 1.1. A Carneades Argument Diagram

6. Label the statements that have been accepted or rejected by the audience to whom the argument was directed.
7. Carneades automatically performs the next step of moving forward over the sequence of argumentation on the basis of which arguments are applicable and which statements have been accepted or rejected to adjust the status of all the statements including the ultimate conclusion.
8. Now the user can critically evaluate the argument by looking to see which conclusions are acceptable in the argument and which (if any) implicit premises have been revealed.

Next let's illustrate how Carneades goes through this procedure by displaying a sequence of argument maps where Carneades has been applied to the example. In Figure 1.1, each premise and conclusion in a typical sequence of argumentation based on argument from expert opinion is shown in a text box. The claim to be proved, the proposition that the proposition *A* may plausibly be taken to be true, is shown in the text box at the far left of the argument diagram. It is the root of the tree. In this instance, the argument from expert opinion is represented as having two premises by combining two of the original premises in the scheme shown at the beginning of section 7. While the premises and conclusions are shown as propositions in text boxes, the arguments are represented as nodes. Where the argumentation scheme can be identified as being of a specific type, its name appears in the argument node. In Figure 1.1 for example, the name of the scheme for argument from expert opinion (EX) is seen within the node at the top. There are four nodes in the tree representing arguments. Each node that contains a plus symbol represents a pro argument that offers positive support, while each node that contains a minus symbol is a contra (con) argument.

So far, this diagram represents the structure of a sequence of argument combining an argument from expert opinion with another argument beneath it that is opposed to the argument from expert opinion. We cannot yet see how to evaluate these arguments as shifts in burden of proof. Looking at Figure 1.1 carefully, we see that in each instance except one, the argument leads by an arrow to a statement that is its conclusion. The exception is the argument that contains the minus symbol in its node. We need to look at this part of the diagram carefully because it shows how the trustworthiness critical question is represented in a special way as an exception premise. This idea needs some careful explanation because there has been a change in how Carneades represents exceptions.

Originally, as stated earlier, Carneades represented the kind of critical question that is an exception as a special type of premise within the argumentation scheme. This method of modeling exceptions has recently been changed however. Instead of modeling an exception as an additional premise, it is now modeled as an undercutter. Pollock (1995, 40) distinguished between two kinds of counterarguments called rebutting defeaters and undercutting defeaters (often referred to as rebutters versus undercutters). A rebutter gives a reason for denying a claim by arguing that the claim is a false previously held belief (Pollock, 1995, 40). An undercutter attacks the inferential link between the claim and the reason supporting it by weakening or removing the reason that supported the claim. A rebutter gives a reason to show the conclusion is false, whereas an undercutter merely raises doubt about whether or not the inference supporting the conclusion holds.

Pollock's red light example (1995, 41) is the best way to explain the distinction.

For instance, suppose *x* looks red to me, but I know that *x* is illuminated by red lights and red lights can make objects look red when they are not. Knowing this defeats the prima facie reason, but it is not a reason for thinking that *x* is *not* red. After all, red objects look red in red light too. This is an *undercutting defeater* (Pollock's italics in both instances).

Normally if I am looking at a light, and it looks red to me, I can assume that it is red. This inference is defeasible, but it does give some reason to draw the conclusion, given the available evidence, that the light I am looking at is red. As Pollock tells us, the inference is indeed defeasible because red lights can make an object look red even when they are not. However, in the absence of the knowledge that there is another red light illuminating the room, which makes the original red light appear to be red, the conclusion can be drawn that the original light is red. However, it is an exception to the rule, if there is a second red light illuminating the room. Hence, the existence of the second red light illuminating the room is an undercutter. It does not defeat the general inference that if I am looking at a light, and it looks red to me I can assume that it is red. This general rule still holds, even though it is subject to exceptions. However, if there is evidence backing up the existence of an exception, the conclusion that the light I am now looking at is red has to be given up.

So now, turning back to Figure 1.1, we can see how the new version of Carneades, instead of treating an exception as a particular kind of premise

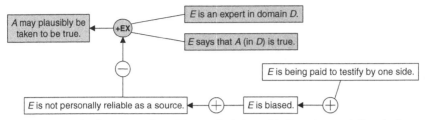

FIGURE 1.2. An Argument from Expert Opinion with Premises and Conclusion Accepted

in an argumentation scheme, treats it as an undercutter. The contra argument from the premise that *E* is not personally reliable as a source undercuts the argument from expert opinion represented in the node described earlier. Notice that this is an instance of a new and special feature of the Carneades Argumentation System. We have an argument shown leading by an arrow to another argument node. This feature represents one argument supporting another argument, or in this instance, it represents one argument undercutting another argument.

Now that we have set up the argument diagram to show how the Carneades Argumentation System works, next we will show how Carneades tracks the shifts in the burden of proof as a critical question is posed.

How this works is illustrated in Figure 1.2, where both the premises of the argument from expert opinion shown at the top right are accepted. Hence, each premise is shown in a darkened box. Given that the argument has the structure of argument from expert opinion, Carneades automatically calculates that the conclusion is accepted as well, and darkens the box in which the conclusion appears. At this point the proponent is winning because his argument from expert opinion proves its conclusion based on the premises that have been offered. Therefore, the burden of proof rests on the opponent who cares to dispute this argument.

Next, we need to ask what happens when the trustworthiness critical question is asked by the opponent. What happens can be illustrated by examining Figure 1.2 once again. The statement "*E* is not personally reliable as a source" is shown as the only premise in a con argument that leads to the node for the argument from expert opinion. But because this premise is an exception, it does not defeat the argument from expert opinion unless it is backed up by additional evidence. Hence, the argument from expert opinion still stands, and its conclusion that *A* may possibly be taken to be true is still shown in a darkened box. Hence, at this point, the burden of proof is still on the opponent because the proponent's argument from expert opinion is still not defeated by the opponent's asking of the critical question. Because this critical question is an exception, it has to be backed up by evidence before it defeats the argument from expert opinion.

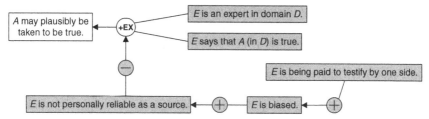

FIGURE 1.3. The Undercutter is Backed Up with Evidence

Next let's turn to Figure 1.3 where some evidence is brought forward that *E* is being paid to testify by one side.

Let's say that the audience accepts this proposition. On this basis, Carneades darkens the text box in which this proposition appears. Then automatically the proposition that *E* is biased will appear in a darkened text box. Because there is now an argument with accepted premises backing up the conclusion that *E* is not personally reliable as a source, now this proposition is shown in a darkened box. Because it is a con argument directed as an undercutter to the argument from expert opinion, the conclusion of the argument from expert opinion is now automatically shown as being placed in a white box, indicating that it is not accepted. Now the burden of proof has shifted against the proponent, who needs to attack the opponent's argument, or find some new argument that can be used to support his ultimate conclusion that *A* may possibly be taken to be true. Unless he can find such additional pro or con arguments, the opponent wins.

What is illustrated even by this simple case is that Carneades has the capability of tracking shifts in the burden of proof from one side to the other during a lengthy sequence of argumentation in which each side takes a turn putting forward opposing arguments and critically questioning the arguments put forward by the other side.

8. Dialogue Models of Argumentation

The most natural setting in which the notion of burden of proof makes sense to us is that of a dialogue in which two parties take turns making speech acts, such as making a claim or putting forward an argument. As the dialogue proceeds, for example, by one party questioning the claim made by the other, the burden of proof is said to shift from one side to the other. During a lengthy dialogue exchange of this sort, the burden of proof may be expected to shift back and forth many times. But this notion of the shifting of burden of proof has always remained to many somewhat shadowy and metaphorical.

The thesis of this book is that the notion of burden of proof can be made sense of in a more precise manner by modeling it in a formal dialogue structure in which two or more parties take turns putting forward speech acts in a connected sequence of moves. Recent work in the area of argumentation studies has already presented us with formal dialogue structures of this type called dialectical models of argumentation. So there are already some resources in place for an investigation of burden of proof. The investigation undertaken in this book is an attempt to see how far such dialectical models of argumentation can take us in yielding a clear and precise analysis of the notion of burden of proof, and to see what needs to be added to them in order to yield such an analysis.

Evaluating argumentation in a dialogue model in which two parties question each other and advance arguments directed to those of the other, is an old idea that goes back to Plato and Aristotle, and even before them to the Sophists. But it wasn't until the research program of the Erlangen School in Germany that efforts were made to carry out a systematic program for constructing formal systems based on the dialogue model. Alexy (1989, 138–154) outlined these historical developments. This research never caught on, however, and was not carried forward. But independently, Charles Hamblin (1970, 1971) constructed mathematical models of dialogue for the practical purpose of providing methods for evaluating fallacies of a kind that had been for a long time of practical interest to logicians. On his model there are three fundamental formal components of a system of dialogue for representing rational argumentation. First, there are two participants (parties), usually called the proponent and the respondent, or White and Black respectively. Second, there is a set of moves made by each party in which the party puts forward a so-called locution, like making an assertion or asking a question. Nowadays these locutions are called speech acts. By convention, White moves first, and then the two take turns. Hamblin (1971, 131–132) showed how these three components are combined to define the concept of a dialogue as an orderly sequence of moves. Such a sequence (Hamblin 1971, 130) is a triple, $<n, p, l>$, where n is a number representing the length of the dialogue (the number of moves so far), p is a participant, and l is a locution. According to Hamblin's definition, a dialogue can be modeled in this way, as a numbered sequence of moves. One type of move is the putting forward of an argument. Thus, an argument can be modeled not only as an inference from a set of premises to a conclusion that has properties like deductive validity and so forth. It can also be viewed as a move that is part of a longer sequence where the conclusion of one argument is reused as a premise in another one.

Another notion that Hamblin introduced is highly characteristic of the types of dialogues he constructed. This is the notion of a commitment set, or so-called commitment store. As each party makes a move, a statement (proposition) is inserted into or deleted from a set of statements attributed to her and kept track of as the dialogue proceeds. A new characteristic of formal dialogues introduced after Hamblin's time is the idea that the sequence of moves in a dialogue moves toward some goal. The idea is that there can be different types of dialogues and each type has a different goal. Hamblin only made a few remarks on such matters. He wrote (1971, 137), that formal systems of dialogue are "information-oriented," meaning that it is assumed that the purpose of the dialogue is the exchange of information among the participants. However, it looks more likely that the formal systems of dialogue constructed in Hamblin (1970) could better be classified as having a goal of rational persuasion of one party by the other through the use of a series of steps of argumentation. In these formal dialogues, the respondent starts out by being uncommitted to some statement, and the efforts of the proponent are directed toward securing the respondent's commitment to this statement through the use of a connected sequence of arguments, one step at a time. It would appear that Hamblin had identified two different types of dialogue, one that seems to have a goal of rational persuasion of one party by the other, while the other type of dialogue is information-oriented. Thus, we are led to the question of whether different types of dialogue can be identified.

Rescher (1977) has a dialogue system containing rules for argumentation that make the type of discussion an asymmetrical one in which a proponent crossexamines a respondent in an adversarial procedure in which the proponent uses the commitments of the respondent as premises to draw conclusions. In this system, the proponent attempts to lead the respondent to either violate some rule of the game or commit himself to an inconsistency. The proponent wins the dialogue by moving the respondent through a series of moves to where he concedes a contradiction. The respondent wins the dialogue if she manages to survive for a designated number of moves without becoming trapped into conceding a contradiction. This type of dialogue will be related to legal examination dialogues in Chapter 5, Section 9.

One aspect of dialectical models of argumentation that may turn out to be an advantage or a disadvantage is their pluralism. For example, a model of burden of proof suitable for legal argumentation might be different in many highly significant respects from a model of burden of proof suitable for everyday conversational argumentation outside a legal setting. It will be the argument of this book that this pluralism is actually an advantage, and even more than that is absolutely necessary to provide us with a clear and precise notion of burden of proof that will be applicable to everyday conversational argumentation.

Six basic types of dialogue have been previously recognized in the argumentation literature (Walton and Krabbe, 1995) and discovery dialogue (McBurney and Parsons, 2001) has been added in to the revised new list of the properties of the basic types of dialogue in Table 1.1. These dialogues are technical artifacts called normative models, meaning that they do not necessarily correspond exactly to real instances of persuasion or negotiation, and so forth, that may occur in a real conversational exchange. Each

TABLE 1.1. *Seven Basic Types of Dialogue*

Type of dialogue	Initial Situation	Goal of Participant	Goal of Dialogue
Persuasion	Conflict of Opinions	Persuade Other Party	Resolve or Clarify Issue
Inquiry	Need to Have Proof	Find and Verify Evidence	Prove (Disprove) Hypothesis
Discovery	Need to Find an Explanation	Find and Defend Hypothesis	Choose Best Hypothesis
Negotiation	Conflict of Interests	Get What You Most Want	Reasonable Settlement
Information-Seeking	Need Information	Acquire or Give Information	Exchange Information
Deliberation	Dilemma or Practical Choice	Co-ordinate Goals & Actions	Decide Best Course of Action
Eristic	Personal Conflict	Verbally Hit Out at Opponent	Reveal Deeper Basis of Conflict

model of dialogue is defined by its initial situation, the participants' individual goals, and the aim of the dialogue as a whole. During the argumentation stage of a dialogue, two parties (in the simplest case) take turns making moves that take the form of speech acts, like asking a question, making an assertion, or putting forward an argument. Dialogue rules define what types of moves are allowed (Walton and Krabbe, 1995). As each party makes a move statements are inserted into or retracted from his/her commitment store. The six basic types of dialogue previously recognized in the argumentation literature are persuasion dialogue, inquiry, negotiation dialogue, information-seeking dialogue, deliberation and eristic dialogue. Discovery dialogue has been added in the new list of the properties of the basic types of dialogue in Table 1.1.

On the account given by McBurney and Parsons (2001, 4), the properties of discovery dialogue and inquiry dialogue are different. In inquiry dialogue, the proposition that is to be proved true is designated at the opening stage, whereas in discovery dialogue the hypotheses to be tested are only formulated during the argumentation stage. A discovery dialogue moves through ten stages (McBurney and Parsons, 2001, 5) called open dialogue, discuss purpose, share knowledge, discuss mechanisms, infer consequences, discuss criteria, assess consequences, discuss tests, propose conclusions and close dialogue.

A dialogue is formally defined as an ordered 3-tuple $\{O, A, C\}$ where O is the opening stage, A is the argumentation stage, and C is the closing stage

(Gordon and Walton, 2009, 5). Dialogue rules (protocols) define what types of moves are allowed by the parties during the argumentation stage (Walton and Krabbe, 1995). At the opening stage, the participants agree to take part in some type of dialogue that has a collective goal. Each party has an individual goal and the dialogue itself has a collective goal. The initial situation is framed at the opening stage, and the dialogue moves through the opening stage toward the closing stage. The type of dialogue, the goal of the dialogue, the initial situation, the participants and the participants' goals are all set at the opening stage. In some instances, a burden of proof, called a global burden of proof, is set at the opening stage, applies through the whole argumentation stage and determines which side was successful or not at the closing stage. In some instances, another kind of burden of proof, called a local burden of proof, applies to some speech acts made in moves during the argumentation stage (Walton, 1988).

Persuasion dialogue is adversarial in that the goal of each party is to win over the other side by finding arguments that defeat its thesis or casts it into doubt. Each party has a commitment set (Hamblin, 1971), and to win, a party must present a chain of argumentation that proves its thesis using only premises that are commitments of the other party. One very well-known type of dialogue that can be classified as a type of persuasion dialogue is the critical discussion (van Eemeren and Grootendorst, 1992). The goal of a critical discussion is to resolve a conflict of opinions by rational argumentation. The critical discussion has procedural rules, but is not a formal model. However, the term 'persuasion dialogue' has now become a technical term of argumentation technology in artificial intelligence and there are formal models representing species of persuasion dialogue (Prakken, 2006).

Inquiry is quite different from persuasion dialogue because it is cooperative in nature, as opposed to persuasion dialogue, which is highly adversarial. The goal of the inquiry, in its paradigm form, is to prove that a statement designated at the opening stage as the *probandum* is true or false, or if neither of these findings can be proved, to prove that there is insufficient evidence to support that the *probandum* is true or false (Walton, 1998, chapter 3). The aim of this type of inquiry is to draw conclusions only from premises that can be firmly accepted as true or false, to prevent the need in the future to have to go back and reopen the inquiry once it has been closed. The most important characteristic of this paradigm of the inquiry as a type of dialogue is the property of cumulativeness (Walton, 1998, 70). To say a dialogue is *cumulative* means that once a statement has been accepted as true at any point in the argumentation stage of the inquiry, that statement must remain true at every point in the inquiry through the argumentation stage until the closing stage is reached. However, this paradigm of inquiry represents only one end of a spectrum where a high standard of proof is appropriate. In other inquiry settings, where there are conflicts of opinion and greater uncertainty, cumulativeness fails, but cooperativeness is a characteristic of inquiry. The model of inquiry dialogue built by Black and Hunter (2009) is meant to represent the cooperative setting of medical domains. Black and Hunter (2009, 174) model two subtypes of inquiry dialogue called argument inquiry dialogues and warrant inquiry dialogues. The former allow agents to share knowledge to jointly construct arguments, whereas the latter allow agents to share knowledge to construct dialectical trees that have an argument at each node in which a child node is a counterargument to its parent.

Inquiry dialogue can be classified as a truth-directed type of dialogue, as opposed to deliberation dialogue, which is not aimed at finding the truth of the matter being discussed, but at arriving at a decision on what to do, where there is a need to take action. While persuasion dialogue is highly adversarial, deliberation is a collaborative type of dialogue in which parties collectively steer actions toward a common goal by agreeing on a proposal that can solve a problem affecting all of the parties concerned, taking all their interests into account. To determine in a particular case whether an argument in a text of discourse can better be seen as part of a persuasion dialogue or a deliberation type of dialogue, one has to arrive at a determination of what the goals of the dialogue and the goals of the participants are supposed to be. Argumentation in deliberation is primarily a matter of identifying proposals and arguments supporting them and finding critiques of other proposals (Walton et al., 2010). Deliberation dialogue is different from negotiation dialogue because the negotiation deals with competing interests, whereas deliberation requires a sacrifice of one's interests.

Deliberation is a collaborative type of dialogue in which parties collectively steer group actions toward a common goal by agreeing on a proposal that can solve a problem affecting all of the parties concerned while taking their interests into account. A key property of deliberation dialogue is that a proposal that is optimal for the group may not be optimal for any individual participant (McBurney et al. 2007, 98). Another property is that a participant in deliberation must be willing to share both her preferences and information with the other participants. This property does not hold in persuasion dialogue, where a participant presents only information that is useful to prove his or her thesis or to disprove the thesis of the opponent. In the formal model of deliberation of McBurney et al. (2007, 100), a deliberation dialogue consists of eight stages: open, inform, propose, consider, revise, recommend, confirm and close. Proposals for action that indicate possible action-options relevant to the governing question are put forward during the propose stage. Commenting on the proposals from various perspectives takes place during the consider stage. At the recommended stage a proposal for action can be recommended for acceptance or nonacceptance by each participant (Walton et al., 2010).

A dialectical shift is said to occur in cases where, during a sequence of argumentation, the participants begin to engage in a different type of dialogue from the one they were initially engaged in (Walton and Krabbe, 1995). In the following classic case (Parsons and Jennings, 1997, 267) often cited as an example, two agents are engaged in deliberation dialogue on how to hang a picture. Engaging in practical reasoning they come to the conclusion they need a hammer and a nail because they have figured out that the best way to hang the picture is on a nail, and the best way to put a nail in the wall is by means of a hammer. One knows where a hammer can be found, and the other has a pretty good idea of where to get a nail. At that point, the two begin to negotiate on who will get the hammer and who will go in search of a nail. In this kind of case, we say that the one dialogue is said to be embedded in the other (Walton and Krabbe, 1995), meaning that the second dialogue fits into the first and helps it along toward achieving its collective goal. In this instance, the shift to the negotiation dialogue is helpful in moving the deliberation dialogue along toward its goal of deciding the best way to hang the picture. For after all, if somebody has to get the hammer and nail, and they can't find anyone who is willing to do these things, they will have to rethink their deliberation on how best to hang the picture. Maybe they will need to phone a handyman, for example. This would mean another shift to an information-seeking dialogue, and involvement of a third party as a source of the information. This example of an embedding contrasts with an example of an illicit dialectical shift when the advent of the second type of dialogue interferes with the progress of the first. For example, let's consider a case in which a union-management negotiation deteriorates into an eristic dialogue in which each side bitterly attacks the other in an antagonistic manner. This kind of shift is not an embedding because quarreling is not only unhelpful to the conduct of the negotiation, but is antithetical to it, and may very well even block it altogether, by leading to a strike, for example.

There are technical problems in building formal models of persuasion dialogue, and the study of these problems have indicated that more than one type of model needs to be considered. The key technical problem in modeling persuasion dialogue is the management of commitment retraction (Walton and Krabbe 1995). Commitments need to be binding, because the whole purpose of the dialogue is to fix commitment. Thus, a participant cannot be allowed to simply retract a commitment at any point, once she begins to run into some sort of trouble attempting to maintain it. On the other hand, commitments cannot be fixed permanently, because tolerance for a certain amount of retraction is necessary. For example, suppose a participant is maintaining a commitment, but it is then shown by the other participant that this commitment is logically inconsistent with some central commitment in the first party's position she had earlier maintained vigorously. Here, the reasonable thing for the participant who has been shown to have inconsistent commitments is to retract one or the other of the commitments that have been cited. The technical problem, as shown by Walton and Krabbe, is to find a system that is loose or flexible enough to allow for such retractions but is also tight enough so that a participant can retract any commitment at any time, even if such a retraction would clearly be at odds with what she is maintaining, or needs to maintain, as her viewpoint in the dialogue.

The solution to this problem worked out by Walton and Krabbe (1995, 126) is to introduce two distinctive types of persuasion dialogue, PPD, or permissive persuasion dialogue, and RPD or rigorous persuasion dialogue. PPD allows for a reasonable freedom of retraction, and also for making of moves, while RPD is much more restrictive in these matters. In a PPD there are two parties, called White and Black, who make moves of various kinds (133). There is an initial conflict description, stating the initial assertions and commitments of both parties (133). Each party has a commitment store (commitment set), a set of statements (134).

Each move of each party can contain various components: retractions, concessions, requests for retractions or concessions, arguments and challenges (135). In a PPD, a move is a six-tuple, and a party can choose to put forward any or all of these six components at each move. Thus there is a good deal of flexibility in how a party make a move in PPD. In contrast, in an RPD, a party can only put forward one of these components at each move. In both PPD and RPD, there are rules governing how the second party must respond to each type of move made by the first party. These rules are rigid in RPD, meaning that only a small number of precisely determine responses are allowed, while in PPD the rules are permissive.

9. Formal Dialogue Models for Legal Argumentation

A natural hypothesis is that such a dialogue model could usefully be applied to legal argumentation. Alexy (1989, orig. German version, 1978) showed how such a dialogue model can be applied to legal argumentation in an influential book. He based his theory on rules of practical discourse governing argument moves made by a proponent and a respondent in legal dialogue. For example, one rule is that a speaker may not contradict himself, while another rule states that whoever has put forward an argument is obliged to defend it. These rules show how Alexy had moved to a dialogue model of legal argumentation, founding a program now being carried forward by a group of researchers in artificial intelligence (AI) and law (Loui, 1998). Recent surveys of this research have shown how important argumentation has become in AI and law (Bench-Capon, 1997). It has also been shown how dialogue-based, or so-called dialectical models, fit well with new developments in AI, especially with new systems used to investigate defeasible argumentation in AI (Hage, 2000).

Another development was the growing interest in legal argumentation on the part of those working in AI. Bench-Capon (1997) recognized the crucial role that argumentation plays in legal justification, and showed how the dialogue format of argumentation is highly compatible with computer formalization. Hage et al. (1994) analyzed procedural reasoning of the kind used in argumentation in so-called hard cases in law. They developed what they call a "dialogical reason-based logic" through the analysis of the reasoning used to justify conclusions in these hard cases. According to their analysis, legal reasoning needs to be seen in a dialogue framework that models an adversarial setting in which there are arguments on both sides of a case. But in order to accommodate their analysis, they showed that just thinking of logical reasoning in law as a chain of inferences is not good enough, and that rules governing the moves of dialogue between the two sides also need to be taken into account. They conclude (113), however, that there is no one set of dialogue rules governing the argumentation in a case: "there are many concurring sets of rules that govern particular types of dialogue." This analysis, along with converging developments in AI and law, pointed the way toward a dialectical treatment of argumentation.

Gordon (1995) developed a dialogue model of legal pleading, the pleadings game, used to identify the legal and factual issues of a case (Gordon, 1995, 109). The pleading stage is the first in a four-stage series of civil proceedings also having a discovery stage, a trial and an appeal stage (110). The plaintiff begins by filing a complaint, and then the defendant may file an answer (111). In the answer, each of the assertions in the complaint can be admitted or denied, or a motion to dismiss can be made (111). The pleadings game analyzes and evaluates legal argumentation within a dialogue format. Lodder (1999) also presented a dialogue model of legal justification that incorporates features of previous dialogue systems, including those of Lorenzen, Barth and Krabbe, Hamblin and Perelman. Lodder's book summarizes many of these earlier systems and comments on how various features of them can be adapted to the study of legal argumentation. It is clear from Lodder's work that the dialogue approach to the analysis of legal argumentation fits the use of AI into modeling legal argumentation.

The model of the persuasion dialogue fits what is often referred to as the advocacy system, held to be the system of dispute resolution in the Anglo-American trial (Frank, 1963). The purpose of the attorney on each side is to win by presenting a more persuasive argument. But the goal of the trial cannot be seen as purely adversarial. It is supposed to provide due process, by having a trier (a judge or jury) who listens to the arguments put forward by both sides, and arrives at a decision by ruling on who presented the more persuasive argument.

The trier resolves the conflict of opinions, not the participants themselves, and the introduction of this third party makes the trial more than just a persuasion dialogue. It is a complex dialogue, with many participants.

The most obvious way a common law trial differs from the model of the persuasion dialogue is that besides the proponent and the respondent, the trial has additional participants such as a judge or a jury who has powers that influence the outcome. These participants can allocate a burden of proof, and can assess probative weight and relevance of arguments, based on rules of evidence and other procedural rules that apply to the argumentation in a trial.

For example, even if one side in a civil case disputes everything the other side says throughout the case, she still loses if the jury or judge finds for the other side. Hence, it is clear that the common law trial is not simply a persuasion dialogue. The relationship between the two can be clarified by recalling Wigmore's distinction (1931) between what he called the science of proof, or principles of logical argumentation generally, and the trial rules used to judge argumentation in a judicial tribunal. Wigmore held that there should be a relationship between these two aspects if the argumentation used in a given trial is to be evaluated as a rational process of drawing a conclusion meant to seek the truth about an issue. Some idea of the complex nature of this relationship is shown in a passage from Wigmore's *Principles* quoted by Twining (1985, 156).

1. That there is a close relation between the Science and the Trial Rules analogous to the relation between the scientific principles of nutrition and digestion and the rules of diet as empirically discovered and practiced by intelligent families.

2. That the Trial Rules are, in a broad sense, founded upon the Science; but that the practical conditions of trials bring into play certain limiting considerations not found in the laboratory pursuit of the Science, and therefore the Rules do not and cannot always coincide with the principles of the Science.

3. That for this reason the principles of the Science as a whole, cannot be expected to replace the Trial Rules; the Rules having their own right to exist independently.

4. But that, for the same reason, the principles of the Science may at certain points confirm the wisdom of the Trial Rules, and may at other points demonstrate the unwisdom of the Rules.

These remarks reveal clearly how the science of rational argumentation in a persuasion dialogue is abstract and general, representing normative rules of a kind that determine the kinds of moves that can be made in rational argumentation. The problem is how such abstract rules apply to individual cases of real argumentation that might occur in an actual trial governed by procedural rules that apply in a given jurisdiction as interpreted by a judge. Such trial rules, as Wigmore observed, have their own right to exist independently as part of an institution. The persuasion dialogue is only an abstract normative (logical) model, whereas an actual trial governed by trial rules is a particular speech event, a case of argumentation used in a social or institutional setting with rules that apply within that jurisdiction.

Another complication is that the persuasion dialogue, at best, only models the argumentation during the argumentation stage. The way evidence enters into a common law trial is a multistaged process. Gordon (1995) has studied the argumentation mainly in the pleadings stage, where the issue is defined. This stage poses the conflict of opinions that is supposed to be resolved by the argumentation that follows. But surrounding this central persuasion dialogue is an elaborate process of dispute resolution and evidence collection in the common law trial that has nine stages (Park et al., 1998, 4–8). First is the pretrial litigation stage, including discovery, motions and hearings. The second stage is jury selection. The third stage is the presentation of opening statements to the assembled court by the attorneys for both sides. At the fourth stage, witnesses are called by the plaintiff, and then examined by both plaintiff and defendant. At the fifth stage, each side has an opportunity for rebuttal. At the sixth stage, either side can make a motion for judgment. The seventh stage is the putting forward of closing arguments by each side that sums up its case. In the eighth stage, the judge instructs the jury on the law that is the basis for deciding the case. In the ninth stage the jury makes its deliberations and reaches a verdict. The common law trial is a complex nesting of dialogues within dialogues, and no formal model of dialogue can encompass all aspects of it. Still, as exponents of the advocacy system have so often maintained, persuasion dialogue has a central place.

Woven around the central persuasion dialogue, however, are not only other embedded types of dialogue, but also procedural rules of various kinds that determine what is allowed into the central persuasion dialogue as evidence. An important factor is that not all arguments are admissible in a common law trial, because rules of evidence lay down requirements on what sorts of arguments can or cannot be presented. These rules of evidence determine what sorts of arguments are admitted and whether they are held to be rationally persuasive. There is also a second type of dialogue embedded in the persuasion dialogue in a trial. It is the information-seeking dialogue, which enables the collection of facts in a case, enabling the argumentation in the persuasion dialogue central to the case to be based on premises that include the relevant information. The persuasion dialogue is most emphasized as a model of argumentation by lawyers who talk or write about trials in the Anglo-American system because the goal of the advocate is to win a case. What may be ignored is that a persuasion dialogue may go off the track and come to a wrong conclusion if the argumentation in it is not based on accurate information that really represents the facts of a case.

This book will move ahead on the hypothesis that persuasion dialogue of some sort is involved centrally in the common law trial, as appears to be suggested by the remarks of those, like Judge Frank (1963), who have defended the common law trial as a method for providing due process on the philosophical basis that it represents an adversarial system in which each side brings out its strongest and most persuasive arguments to clash with those of the other side. Even if this hypothesis is granted, there remain questions on precisely what type of persuasion dialogue it is.

10. A Formal Model of Burden of Proof
in the Critical Discussion

Many readers will be familiar with the critical discussion type of dialogue, which can be classified as a species of persuasion dialogue. Any normative model of dialogue can be defined as an ordered sequence of pairs of moves, which begins at an initial move or opening stage, and proceeds toward a final move, or closing stage. Van Eemeren and Grootendorst (1984) distinguish an initial phase of the critical discussion they identify with the initial confrontation where the participants articulate the goals of the dialogue and clarify or agree on some of the procedural rules. These agreements or clarifications, to the extent that they are known in a particular case, according to an interpretation of the given text of discourse, serve to define the purpose and scope of the dialogue in an overall way. These matters define the context of dialogue in a global manner. Thus, global conventions, rules or agreements pertain to the whole dialogue as a collective sequence of moves. In contrast, local considerations in dialogue pertain to a specific move in the sequence (Walton, 1988).

To begin with, there are two different types of situations that need to be distinguished. In one type of situation, called the dispute (Walton, 2008, 174), the thesis of the one party is the opposite of the thesis of the other. In the other situation, called the dissent in (Walton, 2008, 174), the respondent doubts that the thesis of the proponent is true, and the proponent offers arguments designed to remove that doubt. Hence, every persuasion dialogue can be classified as either a dispute or a dissent, depending on the nature of the initial opposition between the two parties. The dispute can be characterized as an instance of strong opposition, meaning that the thesis of the respondent is the opposite or negation of the thesis of the proponent.

As an example, contrast the dispute between the theist (believer) and the atheist with a dissent between the theist and agnostic. The agnostic claims merely to be skeptical, meaning that he doubts the existence of God, whereas the atheist accepts the proposition that God does not exist. When the respondent is an agnostic, the proponent, that is the theist, merely has to overcome the respondent's expressed doubts, whereas when the respondent is an atheist, the proponent has to not only respond to all the respondent's counterarguments and objections, but also furnish some positive argumentation that supports the claim that God exists.

The dispute is a symmetrical type of dialogue because both parties have a burden of proof to support their thesis. The only difference is that one thesis is the opposite of the other. However, the dissent is an asymmetrical type of dialogue, because only the proponent has the burden of proof. All the respondent needs to do in order to win the argument is to find enough weaknesses in the proponent's argumentation to raise doubts concerning the proponent's claim to have proved his thesis. Previous to this point, there has been no formalized model of the critical discussion, but that has now changed. Krabbe (2013) has built a formal model of the critical discussion (van Eemeren and Grootendorst, 1984, 2004) that is meant to correspond to the argumentation stage and a small part of the opening stage of this type of dialogue (Krabbe, 2013, 8). We will briefly explain here how this model works and how it at least partially represents a notion of the burden of proof operative in a persuasion dialogue.

Van Eemeren and Grootendorst (1983, 82) classified two subtypes of critical discussion. In the simple critical discussion, the protagonist has the task of proving a particular proposition designated at the opening stage as his standpoint, while the antagonist has the job of casting doubt on the protagonist's proof. In the complex critical discussion, both the protagonist and antagonist have a standpoint to be proved, and the standpoint of one is in conflict with that of the other. In the simplest kind of case, the proposition representing the standpoint of the one party could be represented as the negation of the proposition representing the standpoint of the other party. The simple critical discussion corresponds to what is here called the asymmetrical or dissent type of dialogue, while the complex critical discussion corresponds to what we here call the dispute, or symmetrical type of persuasion dialogue.

In the critical discussion of the asymmetrical type, there are two participants called the protagonist and the antagonist, and each has a different role. The protagonist puts forward a thesis at the opening stage and is obliged to defend it during the argumentation stage. The antagonist can challenge the premises of an argument put forward by the protagonist, and she can also challenge the argument itself, that is, the link between premises and conclusion of an argument. This link would generally take the form of a defeasible argumentation scheme, at least in the way we will interpret this type of dialogue here, but it could also take the form of a deductive or an inductive argument. When either party has finished making his or her move, he or she must always put forward the locution, "Your turn."

Winning or losing the critical discussion is determined by defining what counts as the initial standpoint having been successfully defended. Krabbe

(2013, 9) offers the following inductive definition to define this notion: if for any argument that the protagonist has introduced during the course of critical discussion, the link, as well as each of the premises in the argument either remains unchallenged or is supported by the test procedure or by bringing forward arguments to support it, that argument is successfully defended. Readers will need to look carefully at the wording of this inductive definition offered by Krabbe (2013, 9), but the gist of it is conveyed in a way that is easy for the general reader to grasp the general account of it given here. What this means is that either party loses as soon as he or she puts forward the locution "your turn," without responding to the previous move of the other party. In other words, the protagonist wins at the end of the game if the initial standpoint he put forward at the opening stage has been conclusively defended by his moves during the dialogue. Otherwise the antagonist wins.

Here is an example dialogue. At the first move, the protagonist puts forward the statement that Bob shot Ed as his standpoint. At the second move, the antagonist challenged that statement. At the third move, the protagonist puts forward the following argument: witness Wanda said that Bob shot Ed; Wanda was in a position to know that Bob shot Ed; therefore Bob shot Ed. At the fourth move, the antagonist challenged both premises of the argument, and even challenged the argument itself as holding. You might wonder how he could challenge the argument itself, but suppose he said something like this, "witness testimony is not a reliable form of argument." He might be onto something here, for as will be shown in Chapter 4, witness testimony has been shown by social science research to often be unreliable, for example, because memory is fallible, and evidence based on memory is often easily subject to bias. The antagonist said that he doubted whether Wanda really said that Bob shot Ed, he doubted whether Wanda was really in a position to know that Bob shot Ed.

At the fifth move, the protagonist puts forward an argument, and during this same move the two propositions and the argument, the three items that were challenged, are subjected to the testing procedure. Let us consider the outcomes of the testing procedure first. The outcomes of the testing procedure immediately confirm that witness Wanda said that Bob shot Ed by producing a written record in which what Wanda said was directly quoted. The testing procedure also confirms that the argument from witness testimony is a reasonable form of argument that applies to this case. However, the testing procedure fails to confirm the proposition that Wanda was in a position to know that Bob shot Ed. Let us say that there was no knowledge in the database that Wanda was in a position to know about the shooting. Next let us consider the argument that the protagonist put forward at the fifth move. He put forward the argument that a police officer said that Wanda was present at the scene of the killing, and this police officer was also present to observe Wanda being there.

At the sixth move the antagonist challenges the statement that the police officer was in a position to know about Wanda's being at the scene of the killing. At the seventh move, this proposition is tested, and the outcome is that the police officer was there. The protagonist then says "Your turn." However, instead of offering any further challenge, the antagonist merely replies at the eighth move of the dialogue by saying "Your turn."

In Krabbe's formal model, the critical discussion, the argumentation in this example, is shown in Table 1.2 using a key list that represents each proposition.

B: Bob shot Ed.
C: Wanda said that Bob shot Ed.
D: Wanda was in a position to know that Bob shot Ed.
E: A police officer said that Wanda was present at the scene of the killing,
F: This police officer was also present to observe Wanda being there.

The moves in the dialogue are numbered in the first column (M).

In the second column, P stands for the proponent and A stands for the antagonist. As each request to test a proposition or argument is made, the outcome of the test appears in the rightmost column. If a proposition passes the test, it gets an outcome of 1. If it fails the test, it gets a value of 0.

It is shown by the sequence in Table 1.2 that the antagonist has to keep challenging arguments put forward by the protagonist, and that the protagonist, when challenged, has to respond to the challenge by putting forward arguments to defend his claims. So we do see that there is a notion of burden of proof of some sort at work in this formal dialogue system. At move eight, when the antagonist fails to make any new challenge, she immediately loses the dialogue because at each of the previous seven moves the proponent has acted appropriately by either putting forward a standpoint or defending it by means of arguments.

Essentially what burden of proof amounts can be expressed in two different ways for the two types of dialogue. In the asymmetrical type of dialogue, the protagonist has to prove any statement that was challenged. That is his burden. In the symmetrical type of dialogue, the first one to fail to prove any statement that was challenged loses the dialogue. Both parties have the same burden.

However, this formal system does not appear to represent a very robust notion of burden of proof. We can see, for example, that when tested at move 5 the protagonist's proposition D, which was previously challenged by the antagonist, did not get a passing outcome when it was tested. Nevertheless,

TABLE 1.2. *An Example of a Formal Critical Discussion Dialogue*

M	R	Actions at Each Move	Test Result
1	P	Standpoint B: Your Turn	
2	A	Challenge B: Your Turn	
3	P	Argument C,D/B: Your Turn	
4	A	Challenge C: Challenge D: Challenge C,D/B: Your Turn	
5	P	Test C: Test D: Test C,D/B: Argument F,G/D: Your Turn	$C(1), D(0),$ $C,D/B(1)$
6	A	Challenge G: Your Turn	
7	P	Test G: Your Turn	$G(1)$
8	A	Your Turn	

as far as winning or losing the dialogue is concerned, that does not matter. As long as the protagonist responded appropriately by putting forward an argument, the requirements for continuing the dialogue are satisfied. He has not lost the dialogue because he put forward an argument with a premise that did not pass the test.

Still, and for all that, the formal model of critical discussion does display an interesting notion of burden of proof. At each subsequent move, once his standpoint has been put forward as the initial move, the protagonist has to defend against the challenge made by the antagonist. Not only that, the antagonist has a kind of burden as well. She must continue challenging the claims and arguments put forward by the protagonist, or else she will immediately lose the dialogue. We will see in Chapter 2 that there is a type of burden of proof that has been recognized in law that could be modeled by Krabbe's formalization of the asymmetrical type of critical discussion.

The purpose of this book is both to present recent research on burden of proof in AI and law to a wider audience, and to use it to throw new light on how the concepts of burden of presumption and burden of proof work in everyday conversational argumentation. Hence, the line taken in the book goes against the argument that there is no transference from law to everyday conversational reasoning in modeling burden of proof because law has its own rules, argument structures and decision-making procedures that are unique to it. One benefit of this book is that it explains the findings of the leading research on burden of proof and presumption in AI and law in a clear and relatively nontechnical manner, making it useful to anyone interested in legal argumentation. But the broader purpose of the book is to provide a general model of burden of proof and presumption that can be applied to the many other settings where reasoned argumentation is used, including everyday conversational argumentation in public discourse.

Here are the steps we will take along the way toward fulfilling these goals. In Chapter 2 we will explain how the two leading artificial intelligence systems, ASPIC+ and the Carneades Argumentation System, model burden of proof in legal reasoning. In Chapter 8 it will be argued that both of these systems can also usefully be applied to modeling burden of proof in other settings as well, for example, argumentation in a forensic debate, or even everyday conversational argumentation of a kind that is less structured. The main point of Chapter 2 is to get as clear as possible an idea of how these two artificial intelligence argumentation systems work, so that we can then explore the possibilities of not only using them to study interesting facets of legal reasoning, but also to extend them more generally to the study of argumentation in other contexts and fields.

In Chapter 2 it is shown how these developments offer a better way of clearly defining the nature of burden of proof in legal reasoning than was previously possible. Equipped with this precise way of defining burden of proof and its companion notions such as burden of persuasion, we then go on in Chapter 3 to investigate how the notion of presumptions works in legal reasoning within this framework of burden of proof. Chapter 3 will then argue that presumption can be analyzed by showing how it works as a device useful for fulfilling an evidential burden that shifts from side to side in a dialogue. From there Chapter 4 will introduce the reader to the formal dialogue systems that are the backbone of research in argumentation studies and that can model how burden of proof shifts from one side to the other during a sequence of argumentation. What will be fundamentally important here is the idea that argumentation needs to modeled in the context of a dialogue setting that has three basic stages: an opening stage, an argumentation stage and a closing stage. The key to grasping how burden of proof works in such a setting is that there is a global burden of proof set in place at the opening stage that governs how the shifting of an evidential burden from side to side works during the argumentation stage as moves are made by each side. These two stages then determine what the outcome of the dialogue is at the closing stage.

From the point of view of argumentation theory, it has often seemed difficult and problematic to distinguish precisely between the shifting of a presumption and the shifting of the burden of proof in many cases, because both often seem to shift back and forth in a similar way. We might recall that Strong (1992, 425) wrote that presumption is the "slipperiest member of the family of legal terms, except for its first cousin, burden of proof." Thus, the findings of Chapter 2 will work together with the study of the notion of presumption undertaken in Chapter 3 to provide a broader theory explaining how presumption and burden of proof are connected in argumentation. By examining current artificial intelligence models of legal burden of proof and presumption, Chapters 2 and 3 provide the initial foundational steps required for carrying forward a program of research that will fulfill the goal of explaining how burden of proof and presumption are connected at a high level of theory that applies to all logical argumentation.

2

Burdens of Proof in Legal Reasoning

In law, there is a fundamental distinction between two main types of burden of proof (Prakken and Sartor, 2009). One is the setting of the global burden of proof before the trial begins, which is called the burden of persuasion. It does not change during the argumentation stage, and it is the device used to determine which side has won at the closing stage. The other is the local setting of burden of proof at the argumentation stage, often called the burden of production (or the evidential burden, or the burden of going forward with evidence) in law. This burden can shift back and forth as the argumentation proceeds. For example, if one side puts forward a strong argument, the other side must meet the local burden to respond to that argument by criticizing or presenting a counterargument, or otherwise the strong argument will hold, and it will fulfill the burden of persuasion of its proponent unless the respondent puts forward an equally strong objection or counterargument. Otherwise the respondent will lose the trial at that point, and the judge can declare that the trial is over.

According to Williams (2003, 166), considerable confusion has arisen from a failure to distinguish between two distinct kinds of burdens of proof, especially by appeal courts who discuss questions of burden of proof without making it clear whether they are talking about burden of persuasion or evidential burden. Recent ground-breaking work in AI shows great promise for helping law to work toward a more systematic conceptual grasp of the notion of burden of proof by seeing how to model it in a precise way.

Chapter 2 provides a survey showing how the two leading computational models of burden of proof in AI and law apply to legal argumentation by using the running example of self-defense made as a plea in a trial for murder. It is shown how there are three different kinds of burden of proof that need to be distinguished clearly in legal argumentation. By analyzing the self-defense example, Chapter 2 shows how each of these burdens is defined, and how each of them is interlocked with the other to provide an abstract argumentation model that shows how a trial is won or lost by means of the one side or the other producing a sequence of argumentation that fulfills its burden of proof.

1. The Normal Default Rule

The general principle underlying burden of proof in law is often called the normal default rule concerning burden of proof. This rule or principle is expressed in the Latin maxim *necessitas probandi incumbit ei qui agit,* meaning that the necessity of proving falls on the party who acted, i.e., who put forward the action of making the charge, allegation or complaint. One of the most interesting exceptions to the normal default rule, based on considerations of fairness, is that a litigant should not have the burden of establishing facts that are peculiarly within the knowledge of his adversary. The classic case illustrating this exception is that of *United States v. New York 1957.* In this case, the federal government determined that a bill for the services of a carrier had contained certain overcharges, and therefore it deducted these overcharges from the carrier's subsequent bill. The carrier sought recovery of the full amount of the deduction in the United States District Court for the district of Massachusetts. The court ruled that the government had the burden of proving that it was overcharged in the earlier bill. This finding was confirmed by the court of appeals. The case subsequently went to the United States Supreme Court, which reversed the earlier judgment. It ruled that the burden of proving the correctness of the charges disputed by the government remained on the carrier. The rationale offered by the Supreme Court for this ruling was the principle that a plaintiff does not ordinarily have the burden of establishing facts particularly within the knowledge of his adversary (4). The Supreme Court argued that the issue in the case is whether the carrier should have the burden of proving the correctness of the bill or whether the government should have the burden of proving that it was overcharged. Citing previous cases (10), the Supreme Court concluded that the ordinary default rule did not apply in this case because the litigant was not in a position to know about facts that fell within the knowledge of the government.

Another case of this sort was that of *Weast v. Schaffer,* introduced in Chapter 1. This case was brought before the Supreme Court in October 2005. The Individuals with Disabilities Education Act, which requires school districts to create individual education programs for each disabled child, made no statement about the allocation of the burden of persuasion. Hence, the Supreme Court began with the ordinary default rule that the plaintiff bears the burden of proof for the claim made. It was noted that decisions placing the entire burden of persuasion on the opposing party at the outset of the proceeding are extremely rare (2). The most plausible counterargument is that the litigant should not have the burden of establishing facts peculiarly within the knowledge of his adversary

(*United States v. New York*, N. H. & H. R. Co., 355 U.S. 253). This exception to the rule failed in this case because the Individuals with Disabilities Education Act offers parents procedural protections so they can access evidence or expert opinion to match that of the government. It was acknowledged that school districts have a natural advantage in information and expertise. But even so, in this case, it was concluded that the exception to the normal default rule does not apply, because Congress had obliged schools to safeguard the procedural rights of parents and share information with them. According to previous legislation, parents have the right to review all records on their child possessed by a school, and also the right to an independent educational evaluation of their child by an expert. Also, Congress requires school districts to answer the subject matter of a complaint in writing, providing parents with the school's justification for the disputed action. Based on these considerations, the U.S. Supreme Court concluded that the burden of proof was properly placed on the parties seeking relief, that is, the family who argued that an individualized education program should be created for their child.

This case turned out to be a fairly straightforward one. The court did not have to intervene to make the complex rulings on the burden of production during the trial. Instead the issue was one of the allocation of the burden of persuasion. The petitioners argued that putting the burden of persuasion on school districts will help to ensure that children receive a free public education. In terms of argumentation theory, this argument would be classified as argumentation from positive consequences. While it is plausible that assigning the burden of persuasion to schools might have this effect, there is also the practical problem of where the money might come from to pay for the litigation that would result from such a court ruling. There was already concern about the amount of money being spent on the administration of the Individuals with Disabilities Education Program. The Supreme Court argued that this counter failed to provide sufficient justification for departing from the ordinary default rule in this case. As noted earlier, they also reasoned that the more plausible argument for departing from the normal default rule, based on the advantage in knowledge of the government, did not apply in the circumstances of this case.

The normal default rule seems simple enough in the abstract, and it may also seem to be similar to the way burden of proof works in everyday conversational argumentation. However, it will be necessary to distinguish three particular kinds of burden of proof that are recognized in law, the burden of persuasion, the evidentiary burden and the tactical burden of proof (Williams, 2003). Burden of proof as a legal concept needs to be seen as applying centrally to the kind of dialogue called the trial process, although it applies to other areas of law as well like police searches. To try to grasp in general how these three burdens work in law, one needs to see that a trial, like any dialogue process, goes through various stages. There is a pretrial stage where a decision is made to take an action to trial. There is the central trial stage where rules of evidence apply, determining what evidence is admissible and where other procedural rules apply. There is a concluding stage where the trier makes a ruling for one side or the other. Also, there may be posttrial stages that follow, like sentencing or an appeal.

2. Burden of Persuasion and Evidential Burden

According to *McCormick on Evidence* (Strong, 1992, 425), the term "burden of proof" is ambiguous, covering two different notions commonly called burden of proof. The two meanings are often distinguished in law by calling one the burden of persuasion and the other the burden of production. The latter is sometimes also called the burden of producing evidence. The burden of persuasion is described as "the burden of persuading the trier of fact that the alleged fact is true" (425). The place where the burden of persuasion is allocated in a legal proceeding, for instance a trial, is at the point where the judge instructs the trier of fact on what needs to be proved for the issue to be decided. Thus, the burden of persuasion can be described as an obligation that remains on a party to a dispute for the duration of the dispute that once discharged, enables the party to succeed in proving his claim, thereby resolving the dispute. And the most often cited example is the presumption of innocence in criminal law. This burden requires the prosecution to prove all the elements of the offense charged. The prosecution wins the case if and only if it proves all the elements of the charge beyond a reasonable doubt.

Fleming (1961) has carefully drawn the distinction between the risk of nonpersuasion, sometimes called the burden of persuasion, and the duty of producing evidence, sometimes called the burden of going forward with evidence or the production burden. The usual requirement of burden of persuasion in civil cases is that there must be a preponderance of evidence in favor of the party making the claim, that is, the proponent, before he is entitled to a verdict (53). This requirement is usually explained as referring not to the quantity of evidence or the number of witnesses but to the convincing force of the evidence (Fleming, 1961, 53). In criminal cases (54), the burden is to show the guilt of the accused beyond reasonable doubt. This test is very rare as applied to civil cases, but there is an intermediate test (54) that calls for clear and convincing evidence. The burden of production first comes into play at the beginning of the trial. If neither party offers any evidence at the trial, the outcome is that one party will lose. To use Wigmore's phrase, this party may be said to bear the risk of nonproduction of evidence.

Wigmore (1940, 270) drew a distinction between these two meanings of burden of proof. The first one he called the risk of nonpersuasion. Wigmore offered the following example (271) from "practical affairs." Suppose A has a property and he wants to persuade M to invest money in it, while B is opposed to M's investing money in it. A will have the burden of persuasion because unless he persuades M "up to the point of action," he will fail and B will win. Wigmore went on to show how the burden of persuasion works in litigation, in a way similar to that of practical affairs, except that the prerequisites are determined by law (273), and the law divides the procedure into stages (274). The second meaning is called the burden of production. It refers to the quantity of evidence that the judge is satisfied with to be considered by the jury as a reasonable basis for making the verdict in favor of one side (279). If this is not fulfilled, the party in default loses the trial (279). According to Wigmore (284), the practical distinction between these two meanings of burden of proof is this: "The risk of nonpersuasion operates when the case has come into the hands of the jury, while the duty of producing evidence implies a liability to a ruling by the judge disposing of the issue without leaving the question open to the jury's deliberations." Wigmore offered a number of good examples, and went on to discuss shifting of the burden of proof from one side to the other (285). He wrote that the risk of nonpersuasion never shifts, but the duty of producing evidence to satisfy the judge does have this characteristic often referred to as a shifting (285–286).

The burden of persuasion, also often called the risk of nonpersuasion, as noted earlier, remains on the same party for the whole duration of the trial. *McCormick on Evidence* explains how it works (Strong, 1992, 426):

The burden of persuasion becomes a crucial factor only if the parties have sustained their burdens of producing evidence and only when all of the evidence has been introduced. It does not shift from party to party during the course of the trial simply because it need not be allocated until it is time for a decision. When the time for a decision comes, the jury, if there is one, must be instructed how to decide the issue if their minds are left in doubt. The jury must be told that if the party having the burden of persuasion has failed to satisfy that burden, the issue is to be decided against that party. If there is no jury and the judge is in doubt, the issue must be decided against the party having the burden of persuasion.

The burden of persuasion, along with the substantive rules, states what each party has to prove in order to win the trial. It is often expressed in a negative way by calling it the risk of nonpersuasion because it states that a party will lose the trial if she does not produce a strong enough argument to persuade the trier to decide for her side. How persuasive such a winning argument needs to be depends on the standard of proof for that type of trial. In a criminal trial, the prosecution has to prove all the elements of the offense beyond a reasonable doubt. For example, if murder is defined as intentional killing within the jurisdiction, the prosecution has to prove that there was a killing and that the act of killing was intentional. In such a case there are two elements, the act of killing and the intentional nature of the act. This burden of persuasion also requires disproving all the opposing arguments brought forward by the defense. What "proof" or "disproof" means are matters determined by the rules of evidence taken to be binding on the trial.

The two main characteristics of the burden of persuasion as defined earlier are (1) that it is fixed and does not change during the whole trial and (2) that once met, it determines who wins the trial. In the case of a jury trial in criminal law, for example, the judge has the task of explaining the burden of proof to the jury at the beginning of the trial, and the jury has the job of applying this burden to all the evidence that was brought forward during the trial, and to use it to decide the outcome. A third aspect of the burden of persuasion is the standard of proof. It determines how heavy the burden is.

In contrast, the evidential burden changes during the trial, and is often said to shift back and forth from one side to the other (Williams, 1977). The evidential burden, as noted earlier, is also often called the burden of producing evidence or the burden of production. Wigmore summarized the distinction between these two burdens by writing that the risk of non-persuasion operates when the case has come into the hands of the jury, while the evidentiary burden implies a liability to a ruling by the judge disposing of the issue without leaving the question open to the jury's deliberations. What seems to be suggested here is that the evidential burden comes into play during the trial process at some particular point when a particular issue is being discussed and one party fails to produce enough evidence to support his or her side on the issue thereby running the risk of having the trial judge determine that issue in favor of the opponent.

According to Williams (2003, 166) however, the expression evidential burden as commonly used in law, can refer to two distinct notions. In one sense, "evidential burden" means "the burden of producing evidence on an issue on pain of having the trial judge determine that issue in favor of the opponent." In another sense, "evidential burden" means that if the party "does not produce evidence or further evidence he or she runs the risk of ultimately losing on that issue." Williams sees these two notions as distinct because the first involves a question of law while the latter involves a tactical evaluation of who is winning or losing at a particular point during the sequence of argumentation in the trial. He adds that commentators do distinguish between these two concepts, using the expression 'evidential burden' to refer to the former notion, and using the expression 'tactical burden' to refer to the latter notion.

Unlike the burden of persuasion, the burden of producing evidence can shift back and forth from one party to another during the sequence of argumentation in a proceeding (Strong, 1992, 425). Prakken (2001, 261) describes an example from Dutch law that illustrates some key features concerning how burden of proof can shift back and forth during the argumentation stage of a trial:

In Dutch law factual possession of a good creates the presumption that a claimed right to that good (e.g., ownership) indeed exists. Challenging this presumption induces the burden of proving otherwise. Consider now the legal rule that damaging a good constitutes an obligation to pay for the damages to the owner of the good, and consider a case where plaintiff claims that defendant ought to pay for damaging his car, and proves his ownership by pointing at his possession of the car. It now matters at which point defendant directs her attack. If she attacks plaintiff on his subargument concerning ownership, then she has the burden of proving that plaintiff does not own the car. However, if she attacks plaintiff's claim that she has damaged plaintiff's car, it suffices for her to cast doubt, since on this issue the burden of proof rests with the plaintiff.

As Prakken (2001, 9) points out, the example illustrates how the burden of proof can shift even with respect to the same issue. As soon as plaintiff has proven his ownership by showing his possession of the car, the burden of proof that he is not the owner shifts to the defendant.

Prakken (2001, 258–259) describes an example from contract law to show how the components of an ultimate *probandum* can affect how a burden of production shifts back and forth at a local level. It is generally true in legal systems that the proponent who claims the existence of a valid contract has the burden of proving the components of the contract, like offer and acceptance. The party who denies the existence of the contract has the burden of proving why there are exceptional circumstances that prevent the contract from being valid in this case. Prakken (2001, 258) considers an example where plaintiff claims that there is a contract between him and defendant because plaintiff offered defendant to sell her his car and she accepted. In such a case, plaintiff has the burden of proving there was an offer and acceptance, possibly along with some other conditions.

Prakken (2001, 259) extends the example by supposing that plaintiff supports his claim that there was an offer in acceptance by bringing forward witnesses who testify to these facts of offer and acceptance. Defendant then may attack this argument by arguing that the witnesses are unreliable. How would the judge allocate the burden of proof in such a case? Would defendant have to prove his claim that these witnesses are unreliable? Or would plaintiff have to prove his claim that the witnesses are reliable? There are other questions here, as well (Prakken 2001, 259). How strong must defendant's argument be? Prakken (2001, 265–268) constructs a formal system about reasoning concerning burden of proof and applied it to this example. He also considers another aspect of the example by extending it. Suppose defendant, instead of claiming that the witnesses are unreliable, claims that there is an exception because she was insane at the time the contract was allegedly formed. In this instance, the burden to prove her insanity would shift to the defendant.

Without going into the details of this case, or explaining the details of Prakken's formal system for modeling burden of proof in it, it is enough to say here that it is a good example to show, in general, how a burden of proof can shift back and forth at a local level during the argumentation stage in a trial, and how such shifts may require some decision by a judge at that level. In such a case, the initial burden of persuasion may not entirely determine which side has the burden of proof at any given point in the sequence of arguments and counterarguments as the trial proceeds. How such shifts in the burden of proof work, and should be ruled on, are questions that cannot be answered purely by using current nonmonotonic logics according to Prakken (2001, 253). They are what Prakken calls (253) "irreducibly procedural" aspects of defeasible reasoning that can only be modeled by turning to metalevel considerations of the kind studied in Chapter 6. However, even before reaching that point we can use the model of a dialogue with three stages to make sense of how burden of proof can shift during a trial.

According to *McCormick on Evidence* (Strong, 1992, 452–453), the expression presumption of innocence, so widely used to describe legal argumentation, is a misnomer.

Assignments of the burdens of proof prior to trial are not based on presumptions. Before trial no evidence has been introduced from which other facts are to be inferred. The assignment is made on the basis of a rule of substantive law providing that one party or the other ought to have one or both of the burdens with regard to an issue. In some instances, however, these substantive rules are incorrectly referred to as presumptions. The most glaring example of this mislabeling is the 'presumption of innocence' as the phrase is used in criminal cases.

Although courts often insist on the inclusion of this phrase in the charge to the jury, what it really refers to is the prosecution's burden of persuasion. Wigmore made the same point when he wrote (1981, §2511, 530–532), "the 'presumption of innocence' is in truth merely another form of expression for a part of the accepted rule for the burden of proof in criminal cases, i.e., the rule that it is for the prosecution to adduce evidence ..., and to produce persuasion beyond a reasonable doubt." Wigmore did, however, add a subtle point of distinction:

However, in a criminal case the term does convey a special and perhaps useful hint, over and above the other form of the rule about the burden of proof, in that it cautions the jury to put away from their minds all the suspicion that arises from the arrest, the indictment, and the arraignment, and to reach their conclusion solely from the legal evidence adduced. In other words, the rule about burden of proof requires the prosecution by evidence to convince the jury of the accused's guilt: while the presumption of innocence, too, requires this, but conveys for the jury a special and additional caution (which is perhaps only an implied corollary to the other) to consider, in the material for their belief, *nothing but the evidence* i.e., no surmises based on the present situation of the accused. This caution is indeed particularly needed in criminal cases.... So far, then, as the 'presumption of innocence' adds anything, it is particularly a warning not to treat certain things improperly as evidence.

Despite this subtle point, it seems best to treat the expression "presumption of innocence" as not representing a kind of presumption at all, but rather as referring to the burden of persuasion of the kind applicable in criminal cases.

3. Standards of Proof

It is a familiar aspect of burden of persuasion that various different levels are set for successful persuasion, depending on the nature of the dispute that is to be resolved by rational argument. Here we have the familiar standards so often cited in connection with burden of persuasion: scintilla of evidence represents a weak standard, preponderance of evidence a stronger one, clear and convincing evidence still a stronger one and proving something beyond a reasonable doubt represents the highest standard. In a criminal prosecution, the party who has the burden of persuasion of the fact must prove it according to the standard of beyond a reasonable doubt (Strong, 1992, 437). In the general run of issues in civil cases the burden of persuasion is fulfilled by a preponderance of evidence, but in some exceptional civil cases it is fulfilled by clear strong and convincing evidence. There is some controversy about how these standards should be defined precisely. For example, what it means to say that the proof standard is one of preponderance of the evidence (greater weight of the evidence) is open to dispute. According to *McCormick on Evidence* (Strong, 1992, 438) preponderance of evidence means that the argument offered is more convincing to the trier than the opposing evidence. One other standard deserves mention here. Probable cause is a standard of proof used in the United States to determine whether a search is warranted, or whether a grand jury can issue an indictment.

According to *McCormick on Evidence* (Strong, 1992, 447), "The term reasonable doubt is almost incapable of any definition which will add much to what the words themselves imply." When definitions are offered, they are usually expressed in terms of the conviction of the jury, for example to a moral certainty, that the charge against the defendant is true (Garner, 1990, 1380). But formulating the standard of proof in terms of the mental state of conviction of a jury is not very helpful for getting an idea of how this standard should be represented in a normative model of rational argumentation. However one defines the standard, it is taken to be a very high standard based on the notion of reasonable doubt. Courts have held that the legal concept of reasonable doubt itself needs no definition (Strong, 1992, 447), and the reason may be that any carefully crafted definition might have to be so subtle and technical that there would be dangers of misunderstandings if judges were to instruct juries with it. Citing *R v. Ching* (1976 63 Cr. App. Rep. 7 at p. 11), Kiralfy (1987, 15) warned that any attempt to define "reasonable doubt" is "likely to cause more confusion than it dispels." In *R. v. Ching*, the Court of Appeals stated that if judges stopped trying to define "that which is almost impossible to define" there would be fewer appeals. This judicial climate of opinion poses an apparently insurmountable challenge for any attempt to provide a computational model of the beyond reasonable standard. However, as Tillers and Gottfried (2006) have shown, even though there is a well-settled maxim supported by judicial wisdom that the beyond reasonable standard is not quantifiable by assigning probability values to it, it does not follow that this standard of doubt is not open to precise analysis based on a computational argumentation model.

Laudan (2006, 31) has chronicled the history of the beyond reasonable doubt standard, and showed how the attempts to clarify the standard have failed, to the extent that he concludes that a careful reading of the opinions of appellate courts who have sought to clarify the doctrine of beyond a reasonable doubt leads to the conclusion that "not even the keenest minds in the criminal justice system" have been able to agree on a shared understanding of the level of proof needed to convict someone of a crime. He concludes (31) that there is "confusion and lack of clarity" surrounding the beyond reasonable doubt standard in the judicial system. He also shows (2006, 46) that the reason for the confusion is at least partly due to the attempts made by the courts and legal theorists to define this standard using subjective language, including "firm belief in guilt" and "fully convinced." On the other hand, he agrees (45) with the opinion often stated by the courts, including the Supreme Court of Nevada, that the standard of beyond a reasonable doubt is "an inherently qualitative one," and that any attempts to quantify it are more likely to confuse than to clarify matters for a jury. Where does this thorough philosophical critique of the beyond reasonable doubt standard leave us?

It leaves us in a very bad place because the beyond reasonable doubt standard is the tool that judges and juries have to use in order to decide whether a defendant is guilty or not of a criminal offense. We are left without any clear or precise criterion for evaluating whether legal argumentation based on evidence in a trial is successful to fulfill its required burden of proof or not. Because the burden of proof appears not to be defined in a precise way, we are left without a method for evaluating evidential reasoning in such a manner that there is a decision procedure for arriving at the conclusion that one side or the other has won in a criminal trial. How can AI move forward from this apparently very difficult position? In Chapters 5 and 6, it will be shown how there is a way forward by combining argument and explanation.

Farley and Freeman (1995) presented a computational model of dialectical argumentation under conditions where knowledge is incomplete and uncertain. This model has the notion of burden of proof as a key element, where it is defined as the level of support that must be achieved by one side to an argument. According to this account, burden of proof has two functions (156). One is that it acts as a move filter, and the other is that it acts as a termination criterion during argumentation that determines the eventual winner of the dialogue. The move filter function relates to the sequence of intertwined moves put forward by the two parties, often called speech acts, over the sequence of dialectical argumentation. When one party puts forward what Farley and Freeman call an input claim (158), there is a search for support for that claim from the input data. This process has been completed when the claim is supported by propositions from the input data. If no support can be found the argument ends with a loss for the side (158). Thus on their analysis, fulfilling any burden of proof requires at least one supporting argument for an input claim. If side one is able to find support for the claim it made, control passes to the other side, which then tries to refute the argument for the claim using both rebutting or undercutting arguments. If an undercutting move is successful, it may result in a change to the qualification of the claim originally made, or even to the withdrawal of the supporting argument. Put in terms of the theory of van Eemeren and Houtlosser (2002), this back and forth argumentation is characteristic of the speech acts and rejoinders made by both sides during the argumentation stage. The goal of the proponent is to generate the strongest possible arguments for its side, and the goal of the opposing side is to respond to those arguments by making appropriate critical moves, like undercutters and rebuttals.

On the analysis of Farley and Freeman (1995, 160) burden of proof always has two elements: which side of the argument bears the burden, and what level of support is required by that side to fulfill that burden. On their analysis, five different levels of support are recognized in the legal domain in trials (160). The basis of the distinctions drawn between these levels of support is what they call a "defendable argument." They define a defendable argument as one that cannot be defeated with the given input data, noting (160) that this has also been called a plausible argument (Sartor, 1993).

1. The level of support called scintilla of evidence requires at least one weak, defendable argument.
2. The requirement of preponderance of the evidence requires at least one weak, defendable argument that outweighs the other side's arguments.
3. The requirement of a dialectical validity requires at least one credible, defendable argument that defeats all of the other side's arguments.
4. The requirement of beyond a reasonable doubt requires at least one strong, defendable argument that defeats all of the other side's arguments.
5. The requirement of beyond a doubt requires that at least one valid, defendable argument defeats all of the other side's arguments.

On their analysis, burden of proof plays four different roles in argumentation. It is a basis for deciding relevance of a particular argument move, for deciding sufficiency of a move, for declaring when an argument has been concluded and for determining who has won or lost the argument. Their analysis provided a general framework in which burden of proof could be modeled as a dialogue structure containing argumentation and standards of proof. Subsequent research, outlined as follows, showed that using the dialogue structure was the right way to go, but it also showed why standards of proof formulated in the Farley and Freeman model, although they provided a general outline of how to formulate the standards, needed refinement.

Four standards were formally modeled in the Carneades Argumentation System (Gordon and Walton, 2009) in increasing order of the strictness of the standard. As shown in Chapter 1, in the Carneades Argumentation System, there are two sides, and there can be pro- and contra arguments put forward by each side during the argumentation stage. This way of representing the standards of proof can be summarized as follows (Gordon and Walton, 2009).

Scintilla of Evidence

- There is at least one applicable argument.

Preponderance of Evidence

- The scintilla of evidence standard is satisfied, and
- the maximum weight assigned to an applicable proargument is greater than the maximum weight of an applicable contraargument.

Clear and Convincing Evidence

- The preponderance of evidence standard is satisfied,
- the maximum weight of applicable proarguments exceeds some threshold α and
- the difference between the maximum weight of the applicable proarguments and the maximum weight of the applicable contraarguments exceeds some threshold β.

Beyond Reasonable Doubt

- The clear and convincing evidence standard is satisfied and
- the maximum weight of the applicable con arguments is less than some threshold γ.

Notice that on this way of defining the standards of proof, the thresholds α, β, and γ are left open, and is not given fixed numerical values.

These notions of burden of proof have been placed in the setting of a sequential notion of argumentation in a dialogue in the Carneades Argumentation System. A dialogue is formally defined as an ordered three-tuple $\langle O, A, C \rangle$ where O is the opening stage, A is the argumentation stage and C is the closing stage (Gordon and Walton, 2009, 5). Dialogue rules (protocols) define what types of moves are allowed by the parties during the argumentation stage (Walton and Krabbe, 1995). The initial situation is framed at the opening stage, and the dialogue moves through the opening stage toward the closing stage.

These standards can apply to the burden of persuasion set at the opening stage, or to arguments put forward during the argumentation stage. In either instance, note that the burden of persuasion is fixed at the opening stage, in contrast to the local burden, which is only operative during the argumentation stage. The local burden requires that when a participant makes an assertion, or makes a claim of any sort, he is required to give sufficient evidence of the right kind to support the claim, to the appropriate standard of proof. If he fails to fulfill this requirement, the argument is not strong enough to fulfill its required burden.

4. Stages of Dialogue and Legal Burden of Proof

Next we turn to the hypothesis that what is called burden of persuasion in law is a factor that is set at the opening stage of the dialogue in a trial, whereas what is called the burden of producing evidence in law (evidential burden, burden of production) only becomes a factor during the argumentation stage. The argumentation stage is said to occur once the trial procedure is set in motion. The burden of production (burden of producing evidence) is seen as tied to the argumentation stage, and then the outcome of winning or losing a verdict, which occurs at the closing stage, is seen to follow as an outcome of what happens with the burden of production. Once the burden of persuasion is set at the confrontation and opening stages, there is still the question of how it interacts with the burden of production to lead to an outcome in which the prover's task has been fulfilled. It would seem to be a promising hypothesis to conjecture that the burden of persuasion set at the opening stage is the device that enables a decision to be made at the closing stage. It enables this to be done by setting an appropriate standard of proof at the opening stage, and then the argumentation that is put forward during the argumentation stage can be measured by the standard. The problem is to fit the general framework of the stages of dialogue to the problem of defining legal burden of proof so that in the end we can provide a theory that can explain how the various burdens of proof should work in legal argumentation and link to the notion of presumption in legal reasoning.

On this theory, the burden of persuasion in any legal case is set at the opening stage, and the burden of production applies during the argumentation stage. Both burdens then function together at the closing stage to determine the outcome of the argumentation in the whole procedure. This theory can be explained with the help of Figure 2.1.

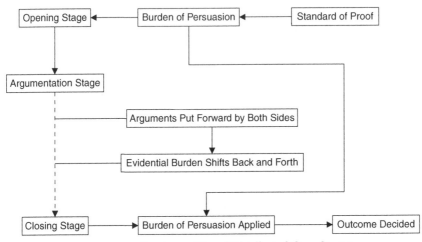

FIGURE 2.1. Burdens of Proof Distributed Over Stages

This theory is completely general, and can apply equally well to a criminal trial, a civil trial or to any other kind of legal dialogue context. It is based on the very important assumption that at the opening stage of any dialogue the proponent must designate a thesis to be proved, and the respondent must agree that the proponent's task has to be fulfilled in order for her to win the dialogue by proving that thesis (Van Eemeren and Houtlosser, 2002). This requirement of rational argumentation by clarification of the issue at the confrontation stage, it will be argued in Chapter 8, can apply to any type of dialogue, legal or otherwise.

We have already seen some indication of it in the formal model of critical discussion of (Krabbe, 2013) described in Chapter 1, Section 10. In this model, once his standpoint has been put forward as the initial move, the proponent has to continually defend it against the challenges put forward by the antagonist. The antagonist must also keep challenging the arguments put forward by the protagonist.

In this model, it is decided what type of procedure the participants will enter into in order to resolve their disagreement at the opening stage. In a legal setting, if the protagonist and the antagonist enter into litigation, they may go forward with a criminal trial or a civil trial. Or they may decide to try to resolve their disagreement by alternative dispute resolution. At this point they are bound to a type of dialogue, and also to a speech event, or a particular institutional framework with procedural rules, for example, a criminal trial in a certain jurisdiction.

The burden of persuasion is set by two factors. At the opening stage, the participants make clear what the issue is. That decides what needs to be proved by each party to secure a favorable outcome at the closing stage.

Here the level of proof required to fulfill the burden of persuasion, and thereby get a winning verdict, is set.

However, in a legal setting, to make the burden of proof shift, it is not enough to put forward any argument supporting your standpoint or most recent argument. The argument must be strong enough. It must be supported by enough evidence to shift the burden. The question is then, how much evidence is required for the proponent to lift this burden. According to Fleming (1961, 55), the general answer to this question can be given in two parts. First, the proponent must put forward sufficient evidence to justify a verdict in his favor. Second, the proponent must put forward sufficient evidence to justify each of the propositions of fact that must be established as part of his case. It is vitally important to note that each thesis will contain beneath it a set of other propositions that are required to be proved in order to prove that thesis. In law, this set of propositions may be called the components or the requirements of the ultimate proposition to be proved, sometimes called the ultimate *probandum*. For example, in order to prove the existence of a contract, the proponent must prove that there was an offer, and acceptance and perhaps other basic requirements that have to be met in a given case before what happened may be properly defined as a contract. The network of argumentation advanced by the proponent to prove all these required components, and the arguments brought against it by the respondent, is put forward at the argumentation stage. This stage can be visualized as a pair of Wigmore diagrams, and also as a sequence of speech acts and rejoinders. Throughout this sequence, the burden of production will shift back and forth, depending on the requirements for a speech act or rejoinder at any given move in the sequence. This mass of evidence as a whole is the basis the trier has for judging at the closing stage whether the proponent's argumentation has fulfilled the burden of persuasion or not, determining the outcome of the trial.

It is important to note that in any type of dialogue, right at the opening stage, not only the ultimate *probandum* will set requirements for burden of proof, but also the other components needed to prove that *probandum*. These requirements are very general, and, we contend, apply to all rational argumentation. As Fleming remarks (1961, 55–56) such general requirements do not get us very far in legal proof by themselves, but simply refer us to two other concepts: the concept of sufficiency of evidence, and the concept of determining what propositions constitute essential elements of a proponent's case. In this chapter, we do not try to say what the burden of persuasion and burden of production should be for each type of trial. That remains to be done. Instead, our theory is a general one that can apply to any argumentation, potentially even one outside the trial context in law, or even outside law altogether. But it is meant to centrally apply to the kind of legal argumentation one would commonly encounter in a trial setting in law.

The problem that remains is that even though the burden of persuasion is set in a trial at the opening stage, that may not be enough by itself to resolve the issue of who won or lost at the closing stage. In some cases, the type of trial, along with the general ruling on burden of persuasion for that type of trial, may be brought to bear on the argumentation put forward by both sides in the trial to clearly determine the outcome of which side won or lost. For example, in a criminal trial, it may be easy for the trier just to decide that the prosecution failed to produce a strong enough argument to prove the guilt of the defendant beyond reasonable doubt. In many cases, the normal default rule that the party that puts forward to claim must prove it can clearly be applied to the case. In such cases, no deep issues about burden of proof arise. These are the easy cases. However, that is not the end of the story. As we will show, burden of proof is the main issue determining the outcome of the trial in some hard cases. As we will show, in these hard cases, each party attempts to shift the burden of proof back onto the side of the other party. To resolve such cases, we will argue, the trier has to probe carefully into the details of burden of production at the argumentation stage. We will argue in Chapter 6 that in these hard cases, the tool needed to provide a rational reconstruction of how the outcome should be decided is that of the burden of proof metadialogue.

There are also other burdens of proof that can be identified in legal argumentation. Prakken and Sartor (2009, 225) explain the difference between burden of persuasion, burden of production and tactical burden of proof as follows. The burden of persuasion specifies which party has to prove its ultimate statement to be proved to the degree required by its proof standard. The failure to prove the statement results in the loss of the proceeding as a whole for that side. The burden of production specifies which party has to offer evidence to support a claim one has made at some particular point in the proceeding. If the evidence put forward does not meet the proof standard for this burden, "the issue is decided as a matter of law against the burdened party, while otherwise the issue is decided in the final stage by the trier of fact according to the burden of persuasion" (2009, 243). Both the burden of persuasion and burden of production are assigned as a matter of law. The tactical burden of proof is not. It is decided by the party himself by assessing the risk of losing on that issue if he presents no further evidence. The tactical burden of proof is fulfilled at a given point during the argumentation stage if, when you add up all your arguments at that point, they are sufficient to fulfill your burden of persuasion. In this chapter, burden of production and tactical burden of proof are subsumed under the general category of local burden of proof.

5. Other Legal Burdens of Proof

Williams (2003, 165) has offered a clearer explanation of the two main ways that the expression burden of proof is commonly used in criminal and civil cases in law. The first meaning is most commonly called the burden of persuasion, but it is also sometimes called the legal burden, the risk of nonpersuasion, the fixed burden of proof and the probative burden. In this sense it refers to "the peculiar duty of him who has the risk of any given proposition on which the parties are issue – who will lose the case if he does not make this proposition out, when all has been said and done" (Thayer, 1898, 355). The second meaning is typically called the evidential burden, but is sometimes also called the burden of production, the burden of producing evidence, the burden of going forward with evidence or the duty of producing evidence satisfactory to the judge.

Williams (2003, 168) argued that the failure to distinguish between two meanings of the expression evidential burden is a subtle source of confusion in law. In one of these meanings, "the evidential burden means the burden of producing evidence on an issue on pain of having the trial judge determine that issue in favor of the opponent." Williams calls this meaning the burden of production, and contrasts it with another type of burden of proof which refers to the burden resting on a party who appears to be at risk of losing on a particular issue at a particular point during the trial: "The party is under an evidential burden and the sense that if the party does not produce evidence or further evidence he or she runs the risk of ultimately losing on that issue." Williams called this burden the tactical burden of proof. According to Williams (2003, 168) ruling on the burden of production involves a question of law, whereas the tactical burden of proof is "merely a tactical evaluation of who is winning at a particular point in time." This distinction between the two senses of the notion of evidential burden is not widely recognized by judges and academic commentators on law. Most legal commentators hold that there are only two types of burden of proof, the burden of persuasion and the evidential burden.

It is commonly said that burden of proof shifts during the course of the trial, but legal commentators generally say that the burden of persuasion is fixed at the opening stage of the trial, never shifts during the trial and is discharged at the closing stage of the trial. Some commentators disagree, however, and say that there are occasions where the burden of persuasion shifts, even though these occasions may be rare. Williams (2003, 168) acknowledged that sometimes it is asserted that both burdens may shift, but he claims that such assertions are based on inadequate analysis of the notion of burden of proof. By his account, the only burden that shifts during the course of the trial is the tactical burden. It is fairly common for influential sources to disagree, however.

The relationship between burden of persuasion and burden of production works in a different way in a criminal case than in a civil case (Prakken and Sartor, 2009, 225–226). In civil cases, the general rule is that the party who makes the claim has the burden of persuasion as well as the burden of production for any claim made, while the other party has both burdens for an exception. For example, in the case of a contract dispute, the party who claims that a contract exists has to prove that there was an offer that was accepted. These are called the two elements of proving a contract. However, there can be exceptions to this rule, for example, the other party might claim that the first party deceived him. In such a case, the party who made the claim that there is a contract has both the burden of production and burden of persuasion for that, while the party who claims that there is an exception has both the burden of persuasion and the burden of production for that. In criminal cases, in contrast, the burden of production and the burden of persuasion can be on different parties. In a criminal case, the prosecution has to meet the standard of beyond reasonable doubt to prove that the defendant is guilty. This principle also covers the nonexistence of exceptions. No weakness in an argument can be left by the prosecution, or proof beyond a reasonable doubt will not be achieved. For example, in a murder case the prosecution has the burden of persuasion to not only prove the two elements (that there was a killing and that it was done with intent), but also to prove the nonexistence of an exception, such as the claim that the killing was done in self-defense. However, the burden of production for proving an exception is on the defense. For example, once the defendant has pleaded self-defense, he will have to provide some evidence to support this claim. Once he has met this burden of production, even by a small amount of evidence, not large enough to meet the requirements of the beyond reasonable doubt standard, the prosecution then has the burden of persuasion that there was no self-defense. It is in this kind of case where the language of shifting the burden of proof is often used to describe the logical mechanism of what has happened. This kind of case is analyzed below in more depth in section 8.

According to Prakken and Sartor (2009, 227), the distinction between the burden of production and tactical burden of proof is usually not clearly made in common law, and is usually not explicitly considered in civil law countries. They add, however, that the distinction is relevant for both systems of law because it is induced by the logic of the reasoning process. Certainly it is not easy at first to grasp clearly the distinction between burden of production and tactical burden of proof, but from the point of view of understanding burden of proof as a concept of logical reasoning, both in law and in everyday conversational argumentation in contexts like philosophical argumentation and political debating, it is highly important to try to do so. The distinction can be clarified by going back to the example of a murder trial where the prosecution has provided evidence to establish killing and intent, and the defense has produced evidence in favor of its plea of self-defense. In such a case, if the prosecution does not rebut the claim of self-defense by producing a counterargument, they stand a very good chance of losing the trial. In such a case, we can say then that the prosecution now not only has the burden of persuasion but also has a tactical burden of proof with respect to the issue of self-defense (Prakken and Sartor, 2009, 227). What is especially interesting is the observation that such a tactical burden can shift back and forth between the parties any number of times during the trial. It depends on "who would be likely to win if no more evidence were provided" (Prakken and Sartor, 2209, 227). To revert to the example, suppose that the prosecution has now provided evidence that goes against the previous argument for self-defense, and the defendant has not rebutted that argument. It is now the defendant who stands to lose. The tactical burden of proof, it can be said, has now shifted to the defendant. However, it is important to note that according to Prakken and Sartor (2009, 227) the burden of production never shifts. Once it has been fulfilled, it is disregarded for the rest of the trial. In their view, the tactical burden is the only one of the three burdens that can be properly said to shift.

The summary distinguishing clearly between the three types of burden of proof offered by Prakken and Sartor (2009, 228) is extremely helpful. Let's consider each of these three burdens in turn. The burden of persuasion specifies which party has to prove some proposition that represents the ultimate *probandum* in the case, and also specifies what proof standard has to be met. The judge is supposed to instruct the jury on what proof standard has to be met and which side has to meet it estimated at the beginning of the trial process. Whether this burden has been met or not is determined at the end of the trial. The burden of persuasion remains the same throughout the trial, once it has been set. It never shifts from the one side to the other during the whole proceedings. The burden of production specifies which party has to offer evidence on some specific issue that arises during a particular point during the argumentation in the trial itself as it proceeds. The burden of production may in many instances only have to meet a low proof standard. If the evidence offered does not meet the standard, the issue can be decided as a matter of law against the burdened party, or decided in the final stage by the trier. Both the burden of persuasion and the burden of production are assigned by law. The tactical burden of proof, on the other hand is decided by the party putting forward an argument at some stage during the proceedings. The arguer must judge the risk of ultimately losing on the particular issue being discussed at that point if he fails to put forward further evidence concerning that issue.

6. The Link between Burden of Persuasion and Production

The big question is how the burden of persuasion at the global level affects the burden of production at the local level. How this works in legal argumentation is somewhat complex, as shown by Prakken and Sartor (2009, 225) because it depends on whether the case is a criminal one or a civil one. In civil cases, the standard of proof is preponderance of evidence, meaning that the plaintiff has to prove her claim on the preponderance of the evidence. For example, to prove that a contract exists between the plaintiff and the defendant, the plaintiff has to prove that she made an offer and that the defendant accepted it. However, there are many exceptions to this rule. For example (Prakken and Sartor, 2009, 226), if one party deceived the other party, or if either party was insane during the time of the offer, these exceptions would refute the claim that there was a contract made. But if the defendant has not produced evidence of his insanity, the plaintiff wins because she will have fulfilled her burden of persuasion according to the standard of the preponderance of evidence. But if she does produce some evidence, then the trier has to judge whether this evidence makes it more probable than not that she was insane. If she did produce any evidence, then she wins the case if the trier finds that it is more probable than not that she was insane. To sum up, in civil cases, the party that makes the claim has both the burden of production and burden of persuasion to prove it, whereas the other party has both the burden of persuasion and the burden of production regarding exceptions.

The linkage between burden of production and burden of persuasion works differently in criminal cases because there the standard of proof, to prove the original claim beyond reasonable doubt, applies not only to the original claim but also to the nonexistence of exceptions (Prakken and Sartor 2009, 226). The reason is that according to the burden of persuasion in a criminal case the accused must be acquitted if there is reasonable doubt. This general principle is also taken to apply to the existence of exceptions. For example, if the defendant pleads self-defense in a murder case, the prosecution has to prove that the killing was not done in self-defense, and this disproof has to meet the reasonable doubt standard. For the burden of production, it is a different matter however. In order to plead self-defense, the accused party has to provide only some evidence to support the claim of self-defense.

It is this kind of case that is frequently described as the shifting back and forth of the burden of proof. First it is on the prosecution. Then it shifts to the defendant who puts up an exception. Next it shifts back to the prosecution to refute the claim made by the defendant. In civil cases in contrast, the plaintiff has both the burden of production and the burden of persuasion to prove her claim, and the defendant has both these burdens for any exceptions that he puts forward. This makes things somewhat complicated from the point of view of trying to use the accepted legal procedures as a way of deriving a model of burden of proof appropriate for argumentation in everyday reasoning, and finding some general way of establishing the linkage between the burden of persuasion and burden of production. On the other hand, it is interesting to see how a difference in the burden of persuasion, with a different standard of proof set at the opening stage of the trial, makes for significant differences in determining how the burden of producing evidence works during the argumentation stage of a trial.

This shifting back and forth of the local burden of proof during an organized disputation has already been described in a general way in the formal dialectical model of Krabbe (2013) described in Chapter 1, Section 10. While this model may turn out to be too simple and limited to offer an analysis of burden of proof in law, it does at least present a clear way of drawing the general distinction between the burden of persuasion and the burden of production of evidence.

At first it seems hard to grasp what the difference is between burden of production and the tactical burden of proof, and generally to understand how the tactical burden of proof works, but an example given by Prakken and Sartor (2009, 227) is helpful. Suppose that in the case of a murder trial the prosecution has provided evidence for killing and intent, and the defense has gone on to produce evidence for self-defense. At this point, the prosecution must assess the risk of losing if they failed to counter the self-defense plea. In such a case, Prakken and Sartor state that the prosecution now has not only the burden of persuasion but also as a tactical burden with regard to the self-defense plea. This tactical burden, as they describe it, can shift back and forth between the parties any number of times during the trial. When it shifts depends on who would be likely to win if no more evidence were provided at that point by the other side. In the example, suppose the prosecution provides evidence to rebut the self-defense plea. Now the burden shifts to the defense, who must estimate the likelihood of their losing if they do not come forward with further evidence supporting their self-defense plea. In such a case it may be said that the tactical burden has now shifted to the defense. Prakken and Sartor (2009, 227) comment that the tactical burden of proof may be contrasted with both burden of persuasion and burden of production because neither shift in the same way that the tactical burden shifts. Once it has been fulfilled, in their opinion, the burden of production never shifts, and is disregarded for the rest of the preceding. This point is an important one, for there is much discussion in the argumentation and philosophy literature about the shifting of the burden of proof. If Prakken and Sartor's view that the burdens of persuasion and production never shift back and forth during the course of the trial is reasonable, it may well be that the copious discussion of the shifting of the burden of proof in the argumentation literature should be taken to refer to the tactical burden of proof.

7. The Abstract Argumentation Model

Prakken and Sartor (2009, 228) have built a logical model of burden of proof in law based on the abstract argumentation framework (Dung, 1995). The proponent starts with the argument he wants to prove and when the opponent has his turn, he must provide a defeating counterargument. An abstract argumentation framework (*AF*) is defined as a pair *Args, Def*, where *Args* is a set of arguments and *Def* $\subseteq$ *Args* × *Args* is a binary relation of defeat. The notion of an argument is taken as primitive, meaning that this model does not reveal anything about the internal construction of an argument (its premises and conclusion, and the kind of inference leading from the premises to the conclusion). The other primitive notion is that of argument defeat. The idea is that each argument can be defeated by other arguments First there is an argument, and then a counterargument that defeats that argument, then a counterargument that defeats that counterargument and so forth, following this kind of sequence.

a2 defeats a1, a3 defeats a2, ..., an defeats an-1

This process is repeated for as long as it takes, and the defeat relationships among a group of arguments can be represented in a simple diagram. An argument is said to be *in* if all its defeaters are out. An argument is *out*, however, if it has a defeater that is in.

To illustrate how argumentation frameworks model a sequence of argumentation, a simple example is presented below. In this example there are six arguments. Each argument is represented by a node, a1, a2, ... a6. The arrows represent defeat relations. For example we see that a2 defeats a1. The argument a6 also defeats a1, but since a1 is already defeated by a2, this additional defeat does not tell us anything new, at least at this stage in the evolution of the argumentation framework. However, it could be significant if additional arguments are taken into account, extending the argumentation framework shown in Figure 2.2.

Another aspect of the argumentation framework in Figure 2.2 that may require some commentary is the relationship between a3 and a4. These two arguments defeat each other. This could represent a stalemate, of a kind sometimes called a zombie argument, in which the two arguments cancel each other out. However, a zombie argument can be resolved if one of the arguments is known to be stronger than the other. For example, if a3 is stronger than a4, then a3 will defeat a4, but a4 will not defeat a3. However, for purposes of this exposition of argument frameworks, we will not take into account how to evaluate the arguments as weaker or stronger. All we need to do is to give a simple outline of how argumentation frameworks provide a formal structure that can be used to model arguments, and defeat relationships among them.

To explain how these frameworks can be applied in a simple case, we extend the example in Figure 2.2 by examining what happens when some

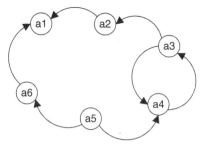

FIGURE 2.2. First Step in the Argumentation Framework Example

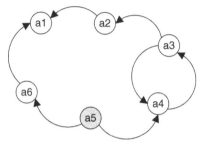

FIGURE 2.3. Second Step in the Argumentation Framework Example

of the arguments are *in* or *out*. In the next sequence of figures, *in* is shown as light gray and *out* is shown as gray. In Figure 2.3, the node a5 is gray showing that it is *in*.

We now need to examine the consequences of this situation are by looking at Figure 2.4. In Figure 2.4, a4 and a6 are shown as dark gray, showing they are *out*, while a5 is shown as light gray.

In Figure 2.4, we can conclude that because a5 is *in*, both a4 and a6 will be *out*. This situation is represented on the diagram by showing both a4 and a6 as dark gray. The reason is that any argument that is defeated by an argument that is *in* becomes an argument that is *out*.

To see what further consequences will follow from this point, we turn to Figure 2.5. Now we see that because a4 is *out*, as shown in Figure 2.5, a3 is not defeated by any other argument that is *in*, a3 is *in*. The arrow going from a3 to a4 now shows redundantly that a4 is *out*, because it is defeated by another argument that is *in*, namely a3.

Next we turn to Figure 2.6, which shows that a2 must be *out*, because it is defeated by an argument that is *in*, namely a3.

At the final step in the sequence of argumentation, we examine what happens with a1, the only argument that is not yet shown to be *in* or *out*. This situation is shown in Figure 2.7. In this last step in the sequence of argumentation, shown in Figure 2.7, node a1 is shown as light gray, showing that argument a1 is *in*. The reasona1 is *in* is that it is not defeated by any

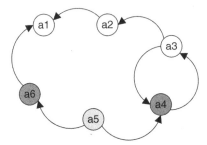

FIGURE 2.4. Third Step in the Argumentation Framework Example

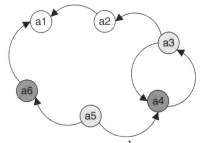

FIGURE 2.5. Fourth Step in the Argumentation Framework Example

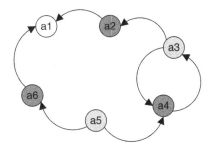

FIGURE 2.6. Fifth Step in the Argumentation Framework Example

arguments that are *in*. It is only attacked by two arguments, namely a2 and a6, and both of these arguments are *out*.

Prakken and Sartor (2006) also argued that presumptions can be modeled as default rules in a nonmonotonic logic. They showed that invoking a presumption can fulfill a burden of production or persuasion while it shifts a tactical version to the other party. Gordon, Prakken and Walton (2007) showed how proof standards make it possible in a formal dialectical system to change the burden of proof during a dialogue as it progresses from stage to stage. Prakken and Sartor (2007) also showed how presumptions are

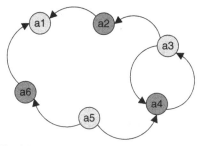

FIGURE 2.7. Final Step in the Argumentation Framework Example

related to burdens of proof. One of their examples, about the contested ownership of archaeological goods in Italy, illustrated how presumption is linked to burden of proof as the dialogue moves forward to its conclusion in a legal case. In a subsequent paper (Prakken and Sartor, 2008), the previous work on presumptions and burden of proof was extended by studying the force of a presumption once counterevidence has been offered.

8. The Self-Defense Example

Prakken and Sartor provide both criminal and civil cases to illustrate their theory of burden of proof, but as interesting as it would be to examine several examples, it is best to choose one for purposes of illustration, so here a criminal case is chosen. This case is an example from Dutch law about murder. As they explain, in criminal cases burdens of production and persuasion on an issue can be on different parties, because proof of guilt by the prosecution requires proof of the nonexistence of exceptions preventing guilt. Self-defense is an exception in a murder case, and so the prosecution has not only the burden of persuasion to prove that the defendant committed murder beyond reasonable doubt, but also has a burden of persuasion to prove there does not exist any exception, like self-defense.

In this regard, the burden of production is different from the burden of persuasion. The prosecution has the burden of production for the so-called elements, which in the example from Dutch law about murder are killing and intent. However, the burden of production is on the defendant to prove exceptions. For example, the burden of production is on the defense to prove self-defense. As Prakken and Sartor (2009, 4) point out "in our murder case example the defence must produce evidence that he [the accused] acted in self-defense but once he has produced such evidence, the prosecution has the burden of persuasion that there was no self-defense." In their analysis of the case, a double arrow is used to represent defeasible rule, a rule that is subject to exceptions. The first rule specifies the elements of the crime of murder according to Dutch law. The second rule specifies that self-defense is recognized by law as an exception to that rule. These two rules are stated as follows (18).

R1: Killing & Intent => Murder
R2: Self-defence => Exception-to-R1

In their example it is supposed that the prosecution can satisfy its burden of persuasion to prove killing and intent by means of evidence. At such a point, the defendant will be convicted if he provides no other evidence, and so the tactical burden shifts to the defense. There is also a burden of production on the defendant, because he can only escape conviction by providing some evidence, even if only a small amount, to the effect that the killing was done in self-defense. In the example, the defense fulfills this burden by providing a witness who claims that the victim threatened the accused with a knife. Rules applicable in this part of the sequence of argumentation are specified as the following three.

F1: Witness W1 says "knife"
R3: Knife => Threat-to-life
R4: Killing & Threat-to-life => Self-defence

Along with these rules, the analysis also uses the argumentation scheme for argument from witness testimony to draw the conclusion "knife" from rule F1.

In the continuation of the example, it is assumed that the defense has satisfied his burden of production for self-defense. As Prakken and Sartor point out, this means that in Anglo-American legal systems the prosecution now needs to provide additional evidence to take away the reasons for doubt raised by this evidence or it will risk losing the trial. So the prosecution has both the burden of persuasion against this claim as well as a tactical burden.

To extend the example further Prakken and Sartor asked us to assume that the prosecution brings forward a witness who declares that the accused had enough time to run away (19). This assumption introduces one more fact and one more rule into their analysis.

F2: Witness W2 says "time-to-run-away"
R5: Knife & Time-to-run-away => ¬Threat-to-life

They also assume (19) that the argument using R5 for ¬Threat-to-life strictly defeats the argument using R4 for Threat-to-life. At this point then, as they observe the proposition "'murder,'" is once again proved, shifting the tactical burden to the defense. For example, the defense could provide some evidence that the witness was a friend of the victim. This would make her testimony unreliable. This development of the example introduces two new rules.

F3: Witness W2 is-friend-of-victim
R5: Witness W is-friend-of-victim => ¬Witness-Testimony-Scheme

R5 represents an undercutter, a kind of counterargument that can defeat the argumentation scheme for the argument from witness testimony. The ultimate effect of these latest arguments is to undercut the prosecution's argument for "¬Threat-to-life," which then reinstates the argument of the defense for "Threat-to-life." This requires a final shift of the burden of proof, one that defeats the prosecution's main claim that the defendant committed murder.

How the sequence of argumentation in this case is modeled by Prakken and Sartor's theory is shown in Figure 2.8, representing the state of the argument at the final move where F3 and R5 are applied. The propositions that are proved by the evidence at that point are shown in the darkened boxes, while the propositions that are not proved are shown in the white boxes. The ultimate conclusion to be proved is shown at the top. The dashed line with the X on it represents defeat.

In Figure 2.8, in argument a1 at the top left, it is shown how the two elements of murder in Dutch law, killing and intent, are taken along with the rule R1 to support the ultimate claim to be proved in this case, the allegation that the defendant committed the crime of murder. However, rule R1 is defeasible. It can default if the exception of self-defense is proved. The defeat relation between "self-defense" and R1 is shown by the dashed line with the X on it joining these two nodes. Notice as well that under the self-defense node, in the argument labeled a2, there are two subarguments. We will combine them both under the label a2. All of the premises in both these arguments are in. This tells us that the argument for "murder" is defeated at this point, given that, taking into account the argumentation at this point, the argument for self-defense (a2) defeats R1, and therefore the argument for "murder" (a1) is defeated.

But now let's look at the rest of the argument (the parts appearing on the left underneath the main argument for "murder"). In this part of the argument, the conclusion that there was no threat to life is supported by the subarguments that appear below the box containing that proposition. This argument contends that the defendant had time to run away, and that therefore there was no threat to his life. This argument would defeat the claim that there was a threat to life, which in turn defeats the argument for self-defense. It is supported by the claim that witness W2 said there was time to run away, shown in the text box at the bottom of Figure 2.8 on the left. This complex we call argument a3. Finally, this witness testimony argument is itself defeated by another argument based on the witness testimony scheme alleging that witness W2 is a friend of the victim. This would mean that witness W2 is biased, and it undercuts the previous argument

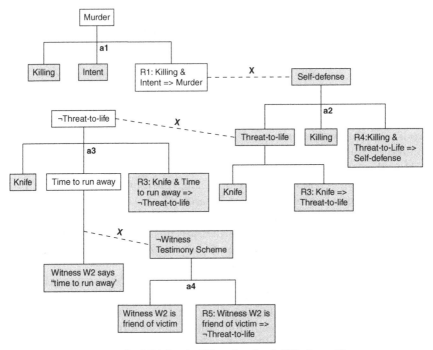

FIGURE 2.8. A Dialectical Tree for the Self-Defense Case

from witness testimony that claimed there was time to run away because witness W2 said so. This is argument a4. By means of a4 the argument a3 concluding that there was no threat to life is defeated, and therefore the argument for self-defense, based on the premise that there was a threat to life, is shown as proved by the evidence. This in turn shows that R1 is not proved, and hence, that the ultimate claim of murder is also shown as not proved.

What Figure 2.8 shows is, that in the example of the Dutch murder case, the burden of production shifts back and forth from one argument to the next, and as we track down the sequence of shifts in the argumentation, ultimately the final shift in the sequence makes the argumentation defeat the claim that the crime of murder can be proved in this case. As each new piece of evidence comes in, it is part of an argument that either supports or defeats a previous argument or claim, making the burden of proof shift from one side to the other. At the end of this procedure in the Dutch murder case, the burden of persuasion to prove the crime of murder is not fulfilled, at the point where all the evidence shown in Figure 2.8 has been brought in and taken into account. This example shows how the burden of persuasion is or is not met at any given point in the sequence of argumentation as evidence is continually being introduced. By tracking through

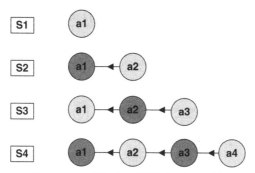

FIGURE 2.9. Sequence of Defeat Shifts in the Self-Defense Case

the support and defeat relations between each of the sub arguments in the sequence of argumentation, we can determine in the given case whether the burden of persuasion is fulfilled or not, as shown in Figure 2.9.

At the beginning argument at step S1, a1 is in, because according to the description of the facts of the case, both killing and intent have been established as facts. Also, rule R1 holds by Dutch law. But at step S2, argument a2 for self-defense is brought in. Argument a2 holds because all its premises, including its two rules, are supported by the facts and the law supposed to apply in the case. Hence, a1 is now defeated. But then once argument a3 is brought in at step S3, argument a2 is defeated. At step S3, because the argument for supporting the claim that there was no threat to life shows that the argument for an exception to the rule R1 is defeated, once again argument a1 is in. Now we proceed to step S4. At this step, argument a3, is undercut by argument a4. Argument a2 is now in, because it is no longer defeated by argument a3. But once argument a2 comes back in, it defeats argument a1.

Understanding how a sequence of argumentation of this sort works shows us how the burden of persuasion is related to the burden of production and tactical burden of proof. It shows how the issue of whether the burden of persuasion has been met can be determined by examining all the connected arguments that support or defeat each other in the mass of evidence leading by a sequence of argumentation into the ultimate claim to be proved.

9. How Carneades Models the Self-Defense Case

In this section, it is shown how Prakken and Sartor's self-defense example can be modeled with respect to the shifts in the burden of proof using the Carneades Argumentation System. We begin with an argument diagram showing how the sequence of argumentation in this example is represented using Carneades. The ultimate conclusion that the defendant committed

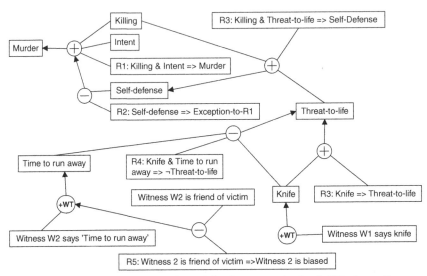

FIGURE 2.10. Carneades Version of the Argument in the Self-Defense Case

murder is shown in the text box at the top and far left of Figure 2.10. This conclusion is supported by a proargument presenting 'killing' and 'intent' as elements of the crime of murder in Dutch law. Also added is the defeasible rule R1 stating that 'killing' and 'intent' defeasibly imply murder. But below that argument there is a contraargument with two premises. These two premises are 'self-defense' and the defeasible rule R2 stating that 'self-defense' is an exception to R1.

To the right of Figure 2.10 at the top there is a proargument supporting the premise stating 'self-defense.' This proargument has three premises. One is threat to life, and another is the rule R3 stating that 'killing' and 'threat-to-life' defeasibly implies 'self-defense.' However, the third premise 'killing' is also required for this argument. What happens here is that the premise 'killing' is reused from the proargument on the left of the one we are currently considering. This illustrates a special feature of Carneades. A premise in one argument can be used again as a premise in another argument. This represents a kind of circular argumentation that one might think should not be permitted in Carneades because the system requires that an argument be modeled as a tree structure, a structure that contains no circles. However, it has been shown that it is technically possible for Carneades to retain this feature of reusing a premise without threatening the consistency and completeness of the system (Brewka and Gordon, 2010).

As we look over the rest of the argumentation shown in Figure 2.10, we see that it represents the remaining arguments in the self-defense example, and especially that it uses the argumentation scheme for the argument from witness testimony in two instances shown at the bottom of the diagram. How the diagram represents the argumentation in the example should now be sufficiently clear to the reader, and we can go on to see how this structure can be used to model the shifts in the burden of proof as each new piece of evidence is brought into the example, according to the description of the example given by Prakken and Sartor.

Carneades has four ways of representing the status of a proposition that is a premise or conclusion in a sequence of argumentation. When an argument is represented on the computer screen using the Carneades Argumentation System, colors and other notations represent the kinds of status that the premises and conclusions have. A proposition can be *stated but not accepted or rejected,* it can be *accepted,* it can be *rejected* or can be *questioned.*

A proposition that has been accepted is shown in a light gray text box containing a checkmark in front of the proposition, whereas a proposition that has been rejected is shown in a dark gray text box containing an X. To make the representation of this example as clear as possible for the reader, however, we will simplify matters by only representing two of these possibilities in the modeling of the example presented as follows. If the proposition is accepted, it will appear in a darkened text box. If the proposition is stated but not accepted, it will appear in a white text box.

In Figure 2.11, the first step of the sequence of argumentation is shown by darkening the four text boxes at the top left of the diagram. What we see visualized here is the first step in the argument, representing the first stage of the introduction of evidence in the example. It is assumed in the example that sufficient evidence has already been given to prove both 'killing' and 'intent,' and that rule R1 applies in the case being considered.

When all three premises are shown as accepted, as visualized in Figure 2.11, the text box containing the ultimate conclusion 'murder' is automatically darkened by the Carneades Argumentation System. This evaluation also depends on the standard of proof that is being used, which in this case, being a criminal case, the standard is that of beyond reasonable doubt. This information can be inserted into Carneades as well, and used as part of the system to model shifting of burden of proof in a given case. However, to keep matters simple for the reader I will not attempt to illustrate this feature in the present example.

According to the account given by Prakken and Sartor of the example summarized in the previous section, the defense fulfilled the burden of production by providing a witness who claimed that the victim threatened the accused with a knife. This second step in the sequence of argumentation is shown in Figure 2.12.

If we look at the bottom right of the argument diagram displayed in Figure 2.12, we see that the defense has brought forward an argument using the scheme for argument from witness testimony based on the premise that witness W1 says there was a knife. This argument defeasibly implies the

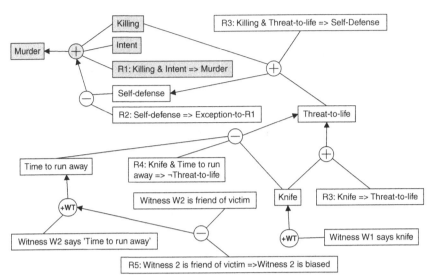

FIGURE 2.11. Step 1 of the Argumentation in the Self-Defense Case

conclusion that there was a knife. Strictly speaking, there should also be another premise to make the argument match the argumentation scheme for argument from witness testimony, but to make the matter as clear and simple as possible for the reader, we have not put this other premise in. However, once the conclusion that there was a knife has been established by the witness testimony argument, it is used in another pro argument along with the additional premise R3, to defeasibly imply the conclusion that there was a threat to life.

Now we need to continue to follow the chain of argumentation leading up the right side of the diagram in Figure 2.12. Once 'threat-to-life' has been established by the witness testimony argument, it can be used in another pro argument, along with the premise 'killing' and the rule R3, to provide an argument for 'self-defense.' Because R3 is a rule of law, it can be treated as accepted, and hence, we have now put it in a darkened box. Notice however that because R3 is accepted as a rule of law, it could also have been put in a darkened box in the previous diagram, Figure 2.11. However, we did not do that because it would have been merely distracting at the earlier point in the discussion when we were considering the argumentation at step 1. However, now we put R3 in a darkened box, showing that it has been accepted, and because the other two premises 'killing' and 'threat-to-life' are also accepted, the conclusion 'self-defense' also appears in a darkened box, showing it has now been accepted, based on the proargument that supports it. At this point, R2 is also shown as accepted, because it is a rule of law, and hence, both premises of the contra argument that undercuts the previous argument from 'killing' and 'intent' to 'murder' are now

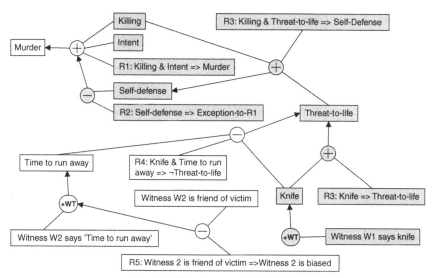

FIGURE 2.12. Step 2 of the Argument in the Self-Defense Case

accepted. In the Carneades Argumentation System, this contra argument is represented as an undercutter, and as a consequence of this, it defeats the argument for 'murder.' So in Figure 2.12, we see an arrow going from the contra argument represented by the node with a minus sign in it to the node above it with the plus sign in it supporting the conclusion that the defendant committed the crime of murder. Hence, Carneades automatically, as shown in Figure 2.12, makes the transition from showing that conclusion in a darkened box in Figure 2.11 to the new result that the same conclusion is shown in a white text box in Figure 2.12. What is shown here is a shift in the burden of production brought about by the new argument based on the witness testimony evidence about the knife.

We now proceed to the next step in the example where the prosecution brings forward a witness who declares that the accused had enough time to run away. This fourth step in the sequence of argumentation is shown in Figure 2.13. To see what has happened at that step we need to look at the bottom left part of the argument diagram shown in Figure 2.13. Witness W2 has testified that there was time for the defendant to run away. Hence based on the defeasible argumentation scheme for argument from testimony, the conclusion 'time-to-run-away' is shown in a darkened text box, meaning that it is now represented as accepted. Now what has happened is that all three premises of the contraargument defeating the claim that there was a threat to life have the accepted status. The premise 'knife' can be reused, because it was already accepted by previous evidence, and the rule R4 can be represented as accepted. We are not told whether this is a rule of in the Dutch case, or whether it is simply a rule that the court

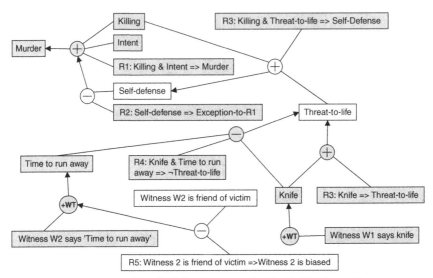

FIGURE 2.13. Step 3 of the Argument in the Self-Defense Case

would accept, but in either event we now show in Figure 2.13 that it is a rule that can be taken to be accepted. Now we have all three premises for the contra argument, so what happens here is that 'threat-to-life,' which was shown in a darkened text box in Figure 2.12, is now shown in Figure 2.13 in a white text box.

Now the Carneades Argumentation System automatically updates the two remaining arguments that depend on this argument. First, it shows 'self-defense' in a white text box, meaning that 'self-defense' is no longer proved, and therefore it shows 'murder' in a darkened text box.

These developments have tracked the shifts in the burden of proof from what was shown in Figure 2.12, where 'murder' was in a white text box and 'self-defense' was in a darkened text box.

Next we proceed to the final development in the argumentation, step 4 shown in Figure 2.14. In this final step, the defense provided some evidence that the witness was a friend of the victim. In the example of Prakken and Sartor, this move was said to be based on a rule to the effect that if the witness is a friend of the victim, this would undercut the argument from witness testimony by making the testimony of the witness appear to be unreliable. Hence, in Figure 2.14, we have represented R5 the bottom in a darkened text box along with the additional premise that witness W2 is a friend of the victim.

What is shown in Figure 2.14 is that the contra argument at the bottom showing the testimony of W2 undercuts the previous argument from witness testimony. So now the proposition that there was time to run away, which had been shown in Figure 2.13 in a darkened box, is shown in a white

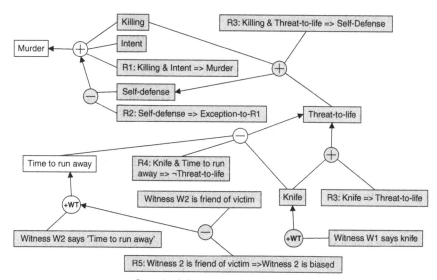

FIGURE 2.14. Step 4 of the Argument in the Self-Defense Case

text box. This reinstates 'threat-to-life,' which can now act as an accepted premise to prove 'self-defense,' and so now the argumentation in favor of murder is once again undercut.

10. Conclusions

It has been shown in this chapter the legal burden of proof can be modeled using two argumentation systems that have been developed in AI. The Prakken and Sartor model of burden of proof is built on the ASPIC+ system of Henry Prakken. The Carneades model is both a theoretical model of argumentation and is also a computational system that has been implemented in the Carneades software tool. In recent research in artificial intelligence these two systems are moving closer together, and indeed it has recently been shown that with regard to their central structures, the two systems are isomorphic with each other (van Gijzel and Prakken, 2011). The Prakken and Sartor system is built on an abstract argumentation framework that portrays burden of proof as changing in a sequence of argumentation as each argument in the sequence defeats the previous argument. The Carneades Argumentation System sees each new argument that is introduced into a case represented by a sequence of argumentation as not only shifting the global burden of proof but also the local burden of proof.

We can also see from the self-defense example that has been analyzed with both systems in this chapter that although the two systems use different technologies, one can see that they both model burden of proof in a roughly comparable and similar away. Although they have different features, it is not hard to see how with some modifications, one system could be adapted to the other in the way it makes compatible determinations on burden of proof in a given case. Hence, the reader can best judge which system appears to be a more useful one. Whichever system turns out to be predominant in the end, both have already proved very useful in offering general structures of argument evaluation that can be very encouragingly applied to modeling the notion of burden of proof. They both show how burden of proof can systematically shift back and forth in a given case as each new item of evidence is introduced in the form of an argument.

One significant problem revealed by Chapter 4 is the issue of what kinds of burdens can shift back and forth. There is central agreement that in legal argumentation the burden of persuasion is set at the opening stage, and represents a global burden of proof that applies over the whole sequence of argumentation, and that is then brought into play at the closing stage to determine which side has won or lost the trial. There is also little doubt that the tactical burden of proof can shift back and forth. However, in Chapter 2 we have frequently described the other burden as shifting back and forth. In law this burden is called the evidential burden, the burden of production, or the risk of nonpersuasion, as shown in section 2. We take it that from the point of view of legal terminology, these terms can all be taken to be equivalent, and refer to the same kind of burden of proof. However, there is a general problem of inconsistency pervading Chapter 2. In this chapter, using intriguing examples, we have frequently described the evidential burden as shifting back and forth at various points in a sequence of argumentation as new arguments are introduced into a case. This way of speaking is very natural, but according to the Prakken and Sartor theory, the burden of production does not shift back and forth during a sequence of argumentation. So here we have a problem to be solved.

One solution is to draw a distinction between legal argumentation and argumentation in other contexts, such as those of a forensic debate or everyday conversational argumentation, by drawing a distinction between and evidential burden that shifts back and forth and a legal burden of production, or risk of nonpersuasion as it can be equivalently called, that does not shift back and forth. Another solution might be to further study how the Prakken and Sartor model and the Carneades model apply to legal argumentation in order to clarify whether the evidential burden really shifts or not, or whether there might be some difference between the burden of production and the evidential burden. These questions for the moment remain open.

3

Presumption in Legal Reasoning

The notions of burden of proof and presumption are central to law, but as we noted in Chapter 1, they are also said to be the slipperiest of any of the family of legal terms employed in legal reasoning. However, as shown in Chapter 2, recent studies of burden of proof and presumption (Prakken, Reed and Walton, 2005; Prakken and Sartor, 2006; Gordon, Prakken and Walton, 2007; Prakken and Sartor, 2007) offer formal models that can render them into precise tools useful for legal reasoning. In this chapter, the various theories and formal models are comparatively evaluated with the aim of working out a more comprehensive theory that can integrate the components of the argumentation structure on which they are based. It is shown that the notion of presumption has both a logical component and a dialectical component, and the new theory of presumption developed in the chapter, called the dialogical theory, combines these two components. Thus, the aim of Chapter 3 is to build on the clarification of the notion of burden of proof achieved in Chapter 2, and to move forward to show how presumption is related to burden of proof. By this means, the goal is to achieve a better theory of presumption.

According to Ashford and Risinger (1969) there is no agreement among legal writers on the questions of exactly what a presumption is and how presumptions operate. However, they think that there is some general agreement on at least a minimal account of what a presumption is: "Most are agreed that a presumption is a legal mechanism which, unless sufficient evidence is introduced to render the presumption inoperative, deems one fact to be true when the truth of another fact has been established" (165). According to legal terminology, the fact to be proved is called "the fact presumed," and the fact to be established before this other fact is to be deemed true is called "the fact proved" (Ashford and Risinger, 1969). The analysis of presumption put forward in this chapter takes this minimal account as its basic structure.

Some of the leading theories in the argumentation and artificial intelligence literature base their analyses of presumption by linking it to the notion of burden of proof as defined in a dialogical model of argumentation that has several stages as a structured, rule-governed dialogue procedure (Prakken and Sartor, 2006, 2007). It is shown in this chapter that the notion of presumption has a logical component and a dialogical component, and the new theory of presumption developed in this chapter, called the dialogical theory, combines these two components. According to the new dialogical theory, in any dialogue there is a burden of proof on either side set at the opening stage of a dialogue, following the account of burden of proof given in Chapter 2. During the argumentation stage, an arguer has an obligation to support any claim made with evidence if she wants the other side to be bound to reasonably accept it. Following the language of Chapter 2, this obligation is called an evidential burden. Unlike the initial burden of persuasion, an evidential burden can shift back and forth during the argumentation stage. Once the argumentation stage is finished, which side won or lost the dialogue is determined by which side put forward a chain of argumentation that met the requirements for its burden of persuasion.

1. The Five Components of Argumentation in a Trial

For this analysis of how burden of proof and presumption work in law, we need the following five components that have been built up so far in the previous two chapters. These components are further elaborated in Chapter 4. The first component is the concept of a trial as a procedure with three main stages, called the opening stage, the argumentation stage and the closing stage. Trials can be divided into various types, but for our purposes it will suffice to distinguish between a criminal trial and a civil trial. In both types of trials, during the argumentation stage two opposed sides present arguments designed to persuade the trier of fact or jury, as we will call it. One side has the goal of proving a claim made at the opening stage, while the other side has the goal of casting doubt on the attempts of the first side to prove its claim. The second component is a set of facts that consists of the evidence judged to be admissible at the opening stage. This set can be added to or deleted from during the argumentation stage. A fact corresponds to what in logic is called a simple proposition, as contrasted with a rule, which takes a conditional form and therefore is classified in logic as a complex proposition. The third component is a set of rules. A rule is set by law previous to the trial and takes the form of a conditional proposition, a complex proposition with an antecedent and a consequent. The fourth component is that of an inference, a structure that has one proposition called the conclusion drawn from a set of other propositions called the premises. For example, from the factual premise "An expert testified to proposition A," and the rule that if an expert testifies to A then

A is acceptable as evidence, the conclusion follows by inference that *A* is acceptable as evidence. This type of evidence is called defeasible because the conclusion may have to withdrawn after the inference has been subject to critical questioning during the argumentation stage.

Verheij (1999, 115) and Walton (2002, 43) have put forward the proposal that many common argumentation schemes fit under a defeasible form of the deductive form of *modus ponens* that we are familiar with in deductive logic. The normal *modus ponens* form of argument that we are familiar with in deductive logic is based on the material conditional binary constant → sometimes called strict implication. The variables *A*, *B*, *C* ... stand for propositions (statements).

Major Premise: $A \rightarrow B$
Minor Premise: A
Conclusion: B

This form of argument can be called strict *modus ponens* (SMP). In contrast, there is also a defeasible *modus ponens* having the following form, where the symbol => is a binary constant representing the defeasible conditional.

Major Premise: $A => B$
Minor Premise: A
Conclusion: B

This form of argument is called defeasible *modus ponens* (DMP) (Walton, 2002, 43).[1] To cite an example, the following argument arguably fits the form of DMP: if something is a bird generally, but subject to exceptions, it flies; Tweety is a bird; therefore Tweety flies. This argument is the canonical example of defeasible reasoning used in computer science. Suppose we find out that Tweety has a broken wing that prevents him from flying, or that Tweety is a penguin, a type of bird that does not fly. If we find out that in the given case one of these characteristics fits Tweety, the original DMP argument defaults. The argument is best seen as not one that is deductively valid, and that still holds even if new information comes in showing that the argument no longer applies to the particular case in the way anticipated. Instead, it is better seen as an argument that holds only tentatively during an investigation, but that can fail to hold any longer if new evidence comes in that cites an exception to the rule specified in the major premise.

Modus ponens arguments, whether of the strict or defeasible type, are typical linked arguments. Both premises go together to support the conclusion. If one is taken away, there is much less support for the conclusion in the absence of the other.

---

[1] Verheij (1999, 115) (2000, 5) called this second form of inference *modus non excipiens*, arguing that it needs to be applied in cases where a general rule admits exceptions.

The fifth component is the process that takes place during the closing stage in which the jury critically examines and evaluates the arguments on both sides that were put forward during the argumentation stage, and decides which one had an argument strong enough to realize its goal of persuasion. The trial is designed so that the argumentation of the one side is opposed to the argumentation of the other side in such a way that if one side fulfills its goal of persuasion, that means that the other side does not.

This normative model of a fair trial is built around concepts drawn from semiformal logic, where an argument has two levels, an inferential level and a dialectical level. At the inferential level, an argument is made up of the inferences from premises to conclusions forming a chain of reasoning. But such a chain of reasoning can be used for different purposes in different types of dialogue. When reasoning is used as argumentation in a dialogue, it needs to be studied at a dialectical level. There are different types of dialogue, and each has its characteristic goals set at the opening stage. One type of dialogue is a persuasion dialogue, in which one party has the goal of persuading the other party to accept a particular proposition called its claim or thesis. The goal of the other party is to resist this attempt at persuasion by expressing critical doubts about the argumentation of the first party. The trial represents a special type of dialogue because it not only has two opposed sides engaging in persuasion dialogue, but it also has other parties that are involved. A judge oversees the trial to see that evidential rules and other rules governing procedure are followed. The decision on which side achieved its goal of persuasion, thereby defeating the other side, may be made by a fourth party, a jury.

On this analysis of the trial the notions of burden of persuasion, evidential burden, tactical burden of proof and presumption can be defined in such a way that we can get a clear account of their relationships to each other as devices useful for the success of a trial. Burden of persuasion is set at the opening stage. It determines which side has to prove what, and what standard of proof is required to prove it, depending on the type of trial. Burden of persuasion does not shift or change during the argumentation stage of the trial. Whether the burden of persuasion has been met by one side or the other is determined by the jury during the closing stage. In contrast, the notions of evidential burden, tactical burden and presumption are operative only during the argumentation stage. To say that there is an evidential burden on a party P concerning proposition A means that A is acceptable as evidence at a particular point during the trial, that A counts against P's side and that if P does not rebut A by presenting evidence to make A no longer acceptable, A will stay in place as acceptable. The difference between evidential burden and tactical burden of proof is a subtle matter discussed in Chapter 2.

A remark of the Florida Supreme Court quoted in (Morman, 2005, 3) brings out the practical nature of presumption very well: "the presumption, shifting the burden of producing evidence, is given life only to equalize the parties' respective positions in regard to the evidence and to allow the plaintiff to proceed." This remark makes several features of the notion of presumption come out. One is that the effect of bringing forward a presumption in a trial is to shift the burden of producing evidence, the evidential burden. A second is that bringing forward a presumption has two purposes. One is to equalize the parties' positions by allowing them to compete on a fairer basis. The other is to allow one of the parties, in this case the plaintiff, to proceed by putting forward his argument without being hampered by a gap in it that is obstructive. It is this gap that makes the respective positions of the two parties unequal.

2. Presumption in Law and Everyday Reasoning

First, let's consider an example from the everyday conversational argumentation. Suppose the chairman of the department sends an email around to all department members describing a proposal that has been forward saying that if she does not get a response from a department member she will assume that any person who does not voice an objection does not object to the proposal and is okay with it. What she has said in the e-mail invokes or puts forward a presumption. The presumption can be seen as a conditional proposition saying that any person who does not put forward an objection to the proposal will be taken not to object to it. This interpretation supports a logical approach. But we can also take a dialectical approach whereby presumption is seen as a kind of speech act that shifts roles and burdens in a dialogue. The chairman is saying that she will presume "no objection" on the ground that she does not hear an objection made in any subsequent message sent by the member this message was sent to. Normally, what would be required to conclude that acceptance of the proposal has been given by a member is a replying e-mail message by the member saying that he or she accepts the proposal. But in this case, because the chairman invokes the notion of a presumption, she is saying that the absence of such a message will be enough to draw the conclusion that the member accepts the proposal (or at least has no objection to it). This effect can be viewed in dialectical terms as a kind of shift of the burden of proof because it places the burden on the member to send a new message to indicate that he or she does not accept the proposal. If the member does not send such a message it will automatically be assumed that he does accept it. This interpretation supports a dialectical approach to presumption, linking it to the shifting back and forth of an existing burden of proof in a structured dialogue setting.

Next let's cite some common examples in law. *McCormick on Evidence*, (Strong, 1992, 455) wrote that there are hundreds of presumptions recognized in law, and here we cite four from his list. (1) A classic instance is the presumption of ownership from possession (454). (2) Presumptions may be made dealing with the survivorship of persons who died in a common disaster, even though there may be no factual basis to believe that one party or the other died first (454). (3) If a first party proves delivery of property to a second party in good condition, and also proves that it was returned in a damaged state, a presumption arises that the damage was due to the second party (456–457). (4) The proof that a person has disappeared from home and been absent for at least seven years, and nobody has heard from this person during that period, raises a presumption that the person died during the seven years (457). Many more examples can be given, but these four are enough to give the reader some idea of how commonly presumptions are used in legal reasoning.

Many presumptions of the kind so often used in law can be classified as evidentiary or probative, but a presumption of consent by silence, of the kind illustrated by the case of the e-mail vote discussed earlier, is based on a constitutive rule. To see this, we need to look at the conditions under which the inference from silence to consent can be defeated. If my side loses the vote, I cannot bring additional evidence that some of those who do not reply to the e-mail did not consent. Under the rules of the discussion, their silence counts as consent. To defeat the presumption, I would have to attack the right of the chair to make this procedural decision, or claim that not everybody received the e-mail. Whether this attack succeeds will depend on the institutional rules. In law, some presumptions are purely constitutive, as in the example of the e-mail case. An example would be the presumption against gift in the law of theft.[2]

To take a closer look at how presumptions work in legal reasoning, consider the following example of another common one (Park, Leonard and Goldberg, 1998, 103).

A presumption states that a letter properly addressed, stamped, and deposited in an appropriate receptacle is presumed to have been received in the ordinary course of the mail. Unless the presumption created by this rule is rebutted, the properly addressed, stamped, and deposited letter will be deemed to have been received in what is considered to be the ordinary amount of time needed in that delivery area.

Notice that the rule stated in the first sentence could be described in logical terms either as a universal generalization or as a conditional statement. On the first description, it says that any letter properly addressed, stamped and deposited in an appropriate receptacle is presumed to have been received. On the second description, it can be expressed as the following

---

[2] I would like to thank an anonymous referee for the journal *Artificial Intelligence and Law* for making this point, and for pointing out as well that different jurisdictions can treat the same rule as constitutive or probative. As this referee pointed out, the rule that a posted letter (see the next example) implies receipt of information is constitutive in German contract law, even though it may well be probative in American law.

conditional: if a letter was properly addressed, stamped and deposited in an appropriate receptacle it will be presumed that it has been received. Notice also that the rule stated in the first sentence of the quotation is also described in that sentence itself as a presumption. This statement suggests the hypothesis that a presumption is the same thing as a general rule of this defeasible kind. But is the rule really the same thing as the presumption, or should a distinction be drawn between the two notions?

The difference between presumption in law and presumption in everyday reasoning outside legal context may be illustrated by some further examples. Take the example of a man who enters a dark room in an old cave and sees something on the floor of the cave that could be a coil of rope or a snake. He has to get past it. On the assumption that it might be a snake, he jumps over it, making sure to clear it as well as possible. In this case, he has acted on the presumption that the object is a snake. The ground for making this presumption is safety. A coil of rope is harmless, but a snake could be dangerous or even deadly. It is a very individual decision whether to jump over the object or not. Jumping over the object could be somewhat dangerous as well, in a cave that is dark and has not been well explored. It is up to the individual to judge what the object looks like in the available light conditions, and to make a personal estimate to what extent it looks more like a snake or more like something harmless.

Contrast this example of a presumption made in everyday reasoning with a typical kind of presumption used in law. Consider once again the example of the legal presumption that if a person has disappeared for seven years, and there is no evidence that he is alive, after this number of years (which varies from jurisdiction to jurisdiction), he is presumed to be dead. As noted earlier, this presumption may be made in law for practical purposes, like that of dividing an estate.

What is the difference between these two kinds of cases? It would appear that the key difference is that the presumption in the snake case is an individualized matter made on the basis of an individual judgment of risk. In the legal case, however, the presumption is a standardized rule of inference accepted by the courts.

3. Rules and Inferences

Some legal writings appear to equate presumption with the rule used to infer a conclusion from a fact, while others define it as the conclusion so drawn. The Florida Evidence Code (F.S. 90.301(2)) defines a presumption as "an assumption of fact which the law makes from the existence of another fact or group of facts found or otherwise established" (Morman, 2005, 3). This definition seems to make the presumption equivalent to the conclusion drawn from the premises of fact and rule. However, another part of the Florida Evidence Code seems to present a different definition: "A presumption of law is a preliminary rule of law which may be made to disappear in the face of rebuttal evidence, but in the absence of such rebuttal evidence, compels a favorable ruling for the party of relying thereupon." This definition appears to equate the presumption with the defeasible rule that is combined with a factual premise to draw a conclusion by inference that can affect the ruling in a trial.

McCormick on Evidence (Strong, 1992, 449) offers the following definition: "a presumption is a standardized practice, under which certain facts are held to call for uniform treatment with respect to their effect as proof of other facts." This definition, at first sight, seems to identify the presumption with a rule on which the inference is based. However, on closer inspection, it identifies the presumption with the standardized practice codified by law in the rule. Presumably, it is the standardized practice that supports the rule, or is the basis of it.

According to this definition, what makes the difference between the kind of presumption found in ordinary reasoning, as in the case of the snake example, and legal presumption, is that legal presumption is a practice that is standardized under court rulings or statutes or both. Yet in both instances, a presumption is something that sanctions an inference from one proposition to another. In the snake case, the inference was drawn from the observation of something that appeared, to some extent, to resemble a snake. From this premise the conclusion was drawn, on grounds of safety, that it would be prudent to jump over the object. In the legal case, the premise was the factual proposition that the person in question had not been seen or heard from for a specified period of time. The conclusion drawn from this premise is that this person can be declared to be dead for legal purposes. The difference between the two is that the legal presumption is a standardized practice set in law as a kind of standard rule that fits all cases meeting the factual specification of the premise. In contrast, an everyday presumption appears to be more of an individual judgment varying with the particulars of the case.

Note that both kinds of presumptions involve inferences from premises to a conclusion, and that in both instances these inferences are defeasible. For example, in the snake case, if the man prods the object with a stick and it does not move, he may reasonably cancel his decision to jump over it in light of this new evidence. The effect of the previous inference is canceled by this new, presented evidence. Similarly, in the legal case, if some new evidence is presented that the man is alive, the conclusion to move ahead on the assumption that he is dead is now revoked.

Ullman-Margalit (1983) recognized that there might be differences between presumptions in law and ordinary conversational reasoning outside a legal framework. She proposed a common underlying framework, however, in the form of a characteristic pattern of inference with three components (1983, 147). The first is the presence of the presumption-raising fact

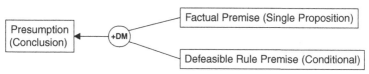

FIGURE 3.1. Presumptive Inference Format

in a particular case at issue. The second is the presumption formula, as she calls it, that sanctions the passage from the presumed fact to a conclusion. The third is a conclusion that a proposition is presumed to be true on the basis of the first two components combined. She described the conclusion of this presumptive inference by saying (147) that the inference is not to a "presumed fact," but to a conclusion that "a certain fact is presumed."

Rescher offers the following defining principle for presumption (2006, 33), where *P* is the proposition representing the presumption: "Any appropriate cognitive presumption either is or instantiates a general rule of procedure of the form that to maintain *P* whenever the condition *C* obtains unless and until the standard default proviso *D* (to the effect that countervailing evidence is at hand) obtains." It is clear that this principle has the form of a qualified conditional of a kind that would be called a defeasible rule in AI.

Based on the approaches of Ullman-Margalit (1983) and Rescher (2006), I propose that a presumptive inference in law or everyday conversational reasoning conforms to the inferential structure shown in Figure 3.1.

In Figure 3.1 we have annotated the linked argument by putting the label *Defeasible Modus Ponens* (DM) on it. As shown in Chapter 3, Section 10, using the scheme for argument from expert opinion as an example, defeasible argumentation schemes of the type that are generally thought to be presumptive in nature can be expressed in the defeasible *modus ponens* form. Thus, the presumptive inference format pictured in Figure 3.1, is meant to be generalized to numerous defeasible argumentation schemes.

The notion of rule is defined according to these characteristics (Gordon, 2008, 4).

1.   Rules have properties, such as their date of enactment, jurisdiction and authority.
2.   When the antecedent of the rule is satisfied by the facts of a case, the conclusion [consequent] of the rule is only presumably true, not necessarily true.
3.   Rules are subject to exceptions.
4.   Rules can conflict.
5.   Some rule conflicts can be resolved using rules about rule priorities, e.g., *lex superior*, which gives priority to the rule from the higher authority.

6. Exclusionary rules provide one way to undercut other rules.
7. Rules can be invalid or become invalid. Deleting invalid rules is not an option when it is necessary to reason retroactively with rules that were valid at various times over a course of events.

The word 'presumably' is used in characteristic 2, indicating the connection between rules and presumptions. As Gordon noted (4), it is a consequence of these characteristics that the notion of a rule cannot be modeled adequately by material implication of the kind we are familiar with in deductive logic. Instead, rules need to be modeled by identifying the parts of the rule, including its antecedent propositions, its consequent proposition, its exceptions, its assumptions and its type.

So far, the analysis only represents the inferential structure of presumption. As we stated in the introduction, a more adequate representation of the notion of presumption has dialogical features as well. To appreciate this, we have to show how the inferential structure shown in Figure 3.1 fits into the five components of argumentation in a trial.

4. The Logical Component

There is now a considerable literature in the field of artificial intelligence and law on the notion of burden of proof (Prakken, Reed and Walton, 2005; Gordon, Prakken and Walton, 2007; Prakken and Sartor, 2007). The notion of burden of proof can be taken as something that has been now fairly well clarified and defined. More recently a theory of presumption has linked the two notions together. The Prakken-Sartor (2006) theory of presumption in law depends on some prior analysis of the notion of burden of proof, and in particular on drawing a three-way distinction among the notions of burden of persuasion, burden of production and tactical burden of proof. According to this theory, presumptions can be characterized as default rules, and disputes about whether a presumption holds do not turn on the question of whether the default rule corresponding to the presumption holds or not.

In order to give a simple example that can be used to illustrate the Prakken-Sartor theory, let's review the case from Chapter 1 Section 5 where the plaintiff demanded compensation on the grounds that the defendant had damaged his bicycle. In this case the plaintiff had the burden of proof for two propositions: that the bicycle was damaged, and that he owned the bicycle. He had both the burden of production and the burden of persuasion to prove these claims. As indicated in the example, the way the plaintiff chose to prove that he owned the bicycle was to claim that he possessed it. This inference is warranted by the general presumption in Dutch law that possession of an object can be inferred from ownership.

The reader will recall that in the Prakken-Sartor theory, a presumption takes the form of a default rule. In this instance the default rule is that normally if a person possesses something, then it can be inferred that he or she owns it. A default rule of this sort holds generally, but is subject to exceptions, and can fail if the other party produces an exception to the rule.

The Prakken-Sartor system is built on a logic for defeasible argumentation called IS (inference system). Information is expressed in the system using a set of rules in the language of extended logic programming. Strict rules are represented with → while defeasible rules are represented with =>. Facts are represented as strict rules with empty antecedents. IS has both negation as failure and classical or strong negation. Arguments can be formed by chaining rules. Conflicts between arguments are decided according to a binary relation of defeat, partly induced by rule priorities, and may be reasoned about like any other issue. Arguments can attack and defeat each other in three ways, and on the basis of such attacks arguments can be assigned dialectical status as winning, losing or tying. Arguments can be "justified" or "defensible."

Prakken and Sartor (2006, 5–6) offer an analysis of the bicycle example designed to show their formal system IS can be applied to the case where a presumption has effects on a burden of persuasion in a trial situation. They begin with two defeasible rules.

```
R1: owner & damaged → compensation
R2: possession → owner
```

Rule R2 expresses a presumption in conditional form, meant to represent the rule in the Dutch civil code that possession of a movable good creates a presumption of ownership. The plaintiff's burden of persuasion amounts to being required to prove the two propositions that he is the owner of the bicycle and that the bicycle was damaged. If he meets these two requirements, the judge will rule that he is to be compensated. Suppose the defendant proves that he possesses the bicycle.

```
F1: possession
```

Now by defeasible *modus ponens* it follows from R2 and F1 that the plaintiff is the owner of the bicycle. Suppose it can be proved that the bicycle was damaged. It follows by R1 that the plaintiff deserves compensation because the burden of persuasion has now been fulfilled by the plaintiff's argument.

What kind of argument could the defendant mount against the plaintiff's argument at this point? Suppose he could provide evidence that the plaintiff had stolen the bicycle in the form of witness testimony to that effect. This argument is represented by Prakken and Sartor in IS by representing it as a fact and two rules.

```
F2:  witness W1 says "stolen"
R5:  witness W1 says "stolen" → stolen
R6:  stolen → ¬owner
```

When taken together, F2, R5 and R6 defeat one conjunct of the antecedent of R1, and this sequence of argumentation now defeats the formerly justified argument that the plaintiff has fulfilled his burden of persuasion. What has happened now is that there has been a shift in the burden of proof. Formerly the plaintiff had fulfilled his burden of persuasion, but with the advent of the new argument, the burden has shifted. Unless the plaintiff can now offer another argument that defeats or brings into question the defendant's argument that he did not own the bicycle because it was stolen, he stands to lose the case.

So described, the bicycle example can be analyzed in terms of the concepts of burden of persuasion and presumption as follows. The plaintiff's case partly depended on the presumption that he owned the bicycle, given the fact that he was in possession of it. This fact is not contested by either side. Because neither side contests the proposition that the defendant damaged the plaintiff's bicycle, it follows by R1 that the plaintiff is entitled to compensation. His burden of persuasion would be fulfilled if the trial were to end at this point. When the defendant offers his counterargument based on witness testimony however, it places an evidential burden on the plaintiff. Unless he can lift this evidential burden, for example, by producing a counterargument to the counterargument, he will no longer have fulfilled his burden of persuasion. He could produce such a counterargument, for example, by arguing that the witness was biased because he was paid to testify for the defendant. But if he fails to make some such move of rebuttal of this sort, his burden of persuasion will no longer be lifted. So one could describe what happened in the case by saying that there has been a shift in the burden of proof. Originally the plaintiff's burden of persuasion was fulfilled, but then, because of the defendant's counterargument, that fulfillment was cancelled.

This way of analyzing the argument in terms of a shift in the burden of proof is, however, not the way that Prakken and Sartor analyze it using their formal system IS. According to their way of analyzing it, the plaintiff has two burdens, both the burden of production and a burden of persuasion, that the bicycle was damaged and that he owned the bicycle (2006, 26). To fulfill this burden the defendant invokes the presumption that the plaintiff owned the bicycle. This presumption is then attacked by the defendant's argument that the plaintiff had stolen it. According to Prakken and Sartor's analysis, "Plaintiff's burden of persuasion for ownership now induces a tactical burden for the plaintiff to convince the judge that he has not stolen the bicycle." Note that they see this move as inducing a tactical burden rather than as a shift in the burden of persuasion or the burden of production.

They explicitly point out that on their analysis, "this is not a burden of production since evidence on the issue of ownership has already been provided, namely possession." To sum up, what has happened in their terms took place in two phases. In the first phase there was a burden of persuasion on the plaintiff's side, one that was met by his initial argument, based on a presumption. In this first phase, the defendant had a tactical burden of producing counter evidence that would introduce reasonable doubt on whether the plaintiff really owned the bicycle. The defendant had fulfilled this tactical burden by producing an argument that the plaintiff had stolen the bicycle, based on witness testimony. The effect of this counterargument was to induce a tactical burden for the plaintiff to convince the judge that he had not stolen the bicycle.

The key to understanding Prakken and Sartor's formal analysis of burden of proof is to see that they stress the importance of the three-way distinction among three kinds of burden of proof in legal argumentation, burden of persuasion, burden of production and tactical burden of proof. As the bicycle example shows, on their analysis presumptions are a way to fulfill a burden of production or a burden of persuasion, and they have the effect of shifting a tactical burden to the other party in a dispute. As this example shows, on their analysis, presumptions can be equated with default rules, and "disputes about what can be presumed should concern whether such a default rule holds" (25). They conclude, "debates about what can be presumed can be modeled in argumentation logic as debates about the backings of default rules" (27). Thus, for Prakken and Sartor, a presumption is always a rule, or what is called a conditional proposition in logic. However, unconditional presumptions, "such as the well-known ones of innocence and good faith," are treated as "boundary cases with tautological antecedents" (23).

5. The Dialogical Component

The three main tasks of argumentation theory are the identification, analysis and evaluation of arguments. Considerable progress has been made in developing methods to assist in the first two tasks through the application of tools like argumentation schemes, profiles of dialogue and argument diagrams. The third task seems more formidable because judging what is a successful or good argument, as opposed to an unsuccessful or bad one, seems to vary so much with different contexts of argument use. For example, a good argument used in the legal context of a trial might not be a good one to be used in a scientific investigation, or vice versa. It might be pretty much the same argument, but the criteria for its acceptance as a good or successful argument might be quite different in the one context than the other. This apparent contextual variability of the task of evaluating arguments is a serious problem, perhaps the main obstacle to moving ahead with a work of developing methods for the evaluation of arguments.

This chapter moves toward a solution to the problem by building on the dialectical approach to argumentation that evaluates an argument not only as a sequence of reasoning formed by chaining of inferences, but also as an entity that occurs in a context of dialogue. Seven basic types of dialogue have been recognized: persuasion dialogue, negotiation, discovery, deliberation, inquiry, information-seeking dialogue and eristic dialogue. The thesis of this chapter is that an argument always needs to be evaluated in relation to the standard of proof appropriate for the type of dialogue in which the argument was put forward. Several different standards of proof recognized in law are distinguished here, and how such standards can be applied to the evaluation of everyday conversational argumentation is studied through the use of some examples. The examples show that solving the problem is only possible if three related but problematic notions, burden of proof, presumptions and argument from ignorance are also taken into account. Luckily, there has been some recent work on the first two notions in argumentation theory that can help us to gain a better understanding of how the last two notions can be analyzed using dialectical models.

Van Eemeren and Houtlosser (2002, 17) argue that burden of proof is necessary in argumentation because it is a procedural concept that is necessary in a critical discussion aimed at resolving a conflict of opinions. On their account, such a conflict can be resolved only if the burden of proof requirement is made clear at the opening stage, and if both parties comply with it throughout the discussion (17). In this way, the concept of burden of proof can be useful as a way of setting the division of labor in a rational discussion. Once the burden of proof is agreed on and set at the opening stage, its effects travel through the whole sequence of argumentation as the participants take turns in the dialogue. For example, the type of speech act called the assertive (where an assertion is made) creates a specific commitment that constitutes a burden of proof because the proponent of an argument is advocating a point of view. The creation of such a burden of proof also creates an obligation to defend an assertion when challenged to do so.

The difference between presumption and burden of proof is that burden of proof is set at the opening stage of the dialogue. On the approach of van Eemeren and Houtlosser, there are four stages of a dialogue in a critical discussion: the confrontation stage, the opening stage, the argumentation stage and closing stage. However, in the past there has been much confusion about the first two stages (Krabbe, 2007). They often get mixed up, and the functions of each of these two stages do not seem to be entirely clear or separate. For this reason in (Walton, 2007) a general theory of the dialogue framework of argumentation was given in which there are always three stages in any type of dialogue – an opening stage, an argumentation stage and closing stage. In this model the first two stages of the Amsterdam model are collapsed into one stage. One reason for adopting this new approach is adapting computer models to argumentation; in any computer model there needs to be three stages, the start point, the endpoint and the sequence of steps joining the start point to the endpoint. Another reason for adopting this approach is that it is applicable to types of dialogue other than just the persuasion dialogue or critical discussion type of dialogue. The initial state for several of these others types of dialogue is not a conflict of opinions posing a confrontation. Instead, there may be a problem to be solved, a proposition to be investigated, or reliable information to be collected, tested and verified. In these types of dialogue there is no confrontation stage.

The three-part distinction (the opening stage, the argumentation stage and closing stage) is vitally important for helping us to understand the difference between burden of proof and presumption. The overarching principle of burden of proof, that he who asserts must prove, is set at the opening stage of the dialogue. Burden of proof is made up at the opening stage by determining three factors: (1) what strength of argument is needed to win the dialogue for a participant at the closing stage (standard of proof), (2) which side bears the so-called burden for producing such an argument and (3) what kind of argument is required for this purpose. "Winning" means producing an argument that is stronger than the opponent's argument to a degree that means that whoever has produced such an argument has succeeded in carrying out the burden of proof set at the opening stage.

When applying this dialogue model from argumentation theory to legal argumentation, another principle is also vitally important: whichever side is in the best position to prove must do so. An example would be jurisdictions that have a reversal of burden of proof in favor of citizens suing the state. Normally the citizen who is suing would have the burden of proof to support his claim, but if he lacks information because that information is possessed only by the government, the burden of proof may shift to the government to provide the required information. This principle is also important as applied to the kind of case in which it is impossible for the defendant to prove absence of a factor. Cases of this sort have been studied in the argumentation literature under the heading of lack of evidence arguments of the type called argument from ignorance (*argumentum ad ignorantiam*) in logic (Walton, 1996).

The key distinction between presumption and burden of proof (of the type called burden of persuasion in law) is that presumption functions only at the argumentation stage, whereas burden of proof is set at the opening stage, has effects at the argumentation stage and is vitally important in determining when the closing stage has been reached in a given case. In a trial, the burden of persuasion is set at the opening stage and remains fixed until the closing stage. At the closing stage, which side won or lost is determined by which side met its burden of persuasion.

There is a distinction always drawn in argumentation theory between two types of persuasion dialogue. In the one type, there is a conflict of opinions, meaning that the one side has a certain thesis, a proposition to be proved, while the other side has the obligation of proving the opposite thesis. This type of conflict of opinions is sometimes called a dispute. In another type, one side has a certain thesis to be proved, but all the other side (the respondent) has to do in order to win the dialogue is to cast enough doubt on the proponent's attempts to prove her thesis such that the proponent is unable to carry out her job of proving it. This type of dialogue is sometimes called a dissent.

The difference between these two types of dialogue is a matter of burden of proof. In the dissent, one side has the burden of proof. The proponent must prove her thesis, by presenting an argument that meets the standard of proof for the dialogue. Otherwise the respondent wins. In the dispute, each side must present its arguments, and the one who has the stronger argument wins. In the dispute, we say that the burden of proof is distributed equally on the two parties. Each side bears a positive burden to prove. Thus, the notion of burden of proof is the fundamental concept that is needed to set up a dialogue by determining at the opening stage what is required on the part of each participant to bring the dialogue to a successful conclusion at the closing stage.

The notion of presumption, in contrast, functions only at the argumentation stage. It functions at individual moves during the argumentation stage. Because of this contrast, it has been said (Walton, 1988) that burden of proof is set at the global level in a dialogue whereas presumption acts at the local level during a sequence of dialogue exchanges. In general, the argumentation stage may be seen as a connected sequence of arguments on each side. You can see it as two lengthy argument diagrams, one representing the chain of argumentation on the proponent's side and the other representing the chain of argumentation on the respondent's side (as indicated in the bicycle example in the previous section). Each party has the job of building up the probative effect of this chain of argument on its own side, as well as the job of criticizing the arguments of the other side and finding the weak points in them. At the closing stage, the two chains of argument are compared. On the basis of the allocation of the burden of proof at the opening stage a decision is arrived at on which side has the winning argument. During any particular point during the sequence of moves on one side or the other, the notion of presumption can come into play. It comes into play when one party or the other makes a claim. The claim is made when the party makes an assertion or otherwise puts forward a particular proposition as something that she claims to be true, or that the other side should accept. The problem is whether the opponent of the claim needs to back it up by an argument, or whether the claim can stand without argument. If a claim made always had to be backed up by an argument, there is the danger of wasting time and energy on claims that nobody would dispute, or perhaps that nobody in the dialogue should have any reason to dispute or right to dispute. For example if the claim is already accepted by all parties to the dispute and nobody has any reason not to accept it, there should be no need to put forward an argument to back it up, even though it might be possible to do so. This is where the notion of presumption is useful. A participant can put forward a claim and ask to have it accepted as a presumption, rather than putting it forward as a flat out assertion. The presumption functions as a reasonable request for the other party to accept the proposition made in the claim without the participant who made the claim having to prove it by offering an argument to support it.

It is during the argumentation stage where arguments are put forward and are critically questioned by the other side where presumptions come into play. When proposition A is put forward by a proponent as a presumption, the productive burden to produce evidence to support that proposition if its acceptance is challenged by the respondent shifts the burden of proof to the respondent's side, but in a negative way. In order not to have to accept A, the respondent must produce evidence against A. In other words, you could say that now the respondent has a burden to rebut the proposition A. This negative logic of presumptions is often called a shift in the burden of proof. However, it is not a shift in the burden of proof that was set at the opening stage of the dialogue. This burden of proof remains constant during the whole dialogue, until the point where it is either fulfilled or not at the closing stage. The shift is one that might be called the productive burden of proof or the tactical burden of proof (Prakken and Sartor, 2007). Such shifts are influenced by argumentation schemes and critical questions.

6. The Letter and the Dark Stairway

Now let's go back to the letter example, cited in Section 2 as a typical case of a legal presumption. The discussion of this example in Section 2 suggested the hypothesis that a presumption is the same thing as a general rule of the defeasible kind, and we can see how this hypothesis is expressed formally by the Prakken-Sartor theory of presumption. However, we also remarked that it can be questioned whether the rule is really the same thing as the presumption, or whether a distinction should be drawn between the two notions. The second sentence of the quotation of the letter example in Section 2 is especially interesting in this regard because it says that the presumption is created by the rule. This wording suggests that the presumption may be something different from the rule. What is perhaps suggested is the following structure. There is a generally accepted rule in law that if a letter was properly addressed, stamped and deposited in an appropriate receptacle, it can be presumed to have been received in the ordinary course of the mail. As noted earlier, this rule from a logical point of view has a conditional form, a form of a qualified universal generalization, of the kind that makes it easily fit with other statements in familiar sequences of logical reasoning. For example, when joined with a particular statement of fact, it can create a logically valid inference. Suppose that in a particular case it has been factually established by evidence that this particular letter was properly addressed, stamped, and deposited in an appropriate receptacle. Using this statement as the antecedent of the *modus ponens* inference along with the conditional statement in the general rule above, the conclusion can be drawn on a defeasible basis that this particular letter was received. The form of inference here is that of defeasible *modus ponens*. In this analysis, the presumption can be seen as something different from the rule. The rule is the conditional statement in the defeasible *modus ponens* inference. The presumption is the statement that this particular letter has been received. That is the presumption created in this case by the application of the rule to the facts of the case.

According to this account of the reasoning, the presumption is not the general rule that is applied to the particular case, that a letter properly addressed, stamped and deposited in an appropriate receptacle can be taken to have been received in the ordinary course of the mail. The presumption is the statement that a particular letter in a particular case has been received. Or perhaps it is even more accurate to say that it can be presumed that a conclusion can be drawn because it can be presumed that the person in a particular case received it. As *McCormick on Evidence* (Strong, 1992) warned, the notion of presumption is slippery and vague. Which meaning should we choose? What should we say? Is the conditional rule the presumption? Is the particular statement inferred from the rule in this particular case the presumption? Or is the whole network of reasoning whereby the particular statement is derived inferentially from the rule and the fact the presumption? If the last option is the best one, it makes sense to define presumption in terms of the sequence of presumptive reasoning whereby an inference is used to draw a presumptive conclusion from a defeasible conditional and its antecedent. These questions can be answered by extending the letter example into a case from civil law.

In the following example of a legal case, which we will call the dark stairway example, summarized from Park, Leonard and Goldberg (1998, 103), the argumentation turns on the presumption stating that a letter properly addressed, stamped and deposited in an appropriate receptacle is presumed to have been received in the ordinary course of the mail.

The dark stairway example is summarized below (from Park, Leonard and Goldberg, 1998, 107):

The plaintiff suffered a fall on a dark stairway in an apartment building. She sued the defendant, the building's owner, claiming that he did not keep the stairway in a safe condition, because the lighting did not work properly. To prove notice, the defendant claimed he mailed a letter to the plaintiff, informing her that several of the lights in the stairway no longer worked.

To see how the letter delivery presumption figures in the chain of reasoning in the case, let us represent the rules and facts in the arguments on both sides using the IS model of Prakken and Sartor.

Rules and Facts in Plaintiff's Argument

```
R1: ¬lighting works properly → ¬stairway in a safe
    condition
F1: ¬lighting works properly
R2: ¬stairway in a safe condition → owner liable
```

Rules and Facts in Defendant's Argument

```
F2: defendant mailed letter to plaintiff, saying
    lights in the stairway don't work
R3: defendant mailed letter to plaintiff, saying
    lights in the stairway don't work → plaintiff was
    informed ¬stairway in a safe condition
R4: plaintiff was informed ¬stairway in a safe con-
    dition → ¬owner liable
```

The chain of reasoning in the plaintiff's argument works as follows. By R1 and F1, based on DMP (defeasible *modus ponens*), the conclusion ¬stairway in a safe condition follows. By this conclusion and R2, the conclusion that the owner is liable follows by DMP. The chain of reasoning in the defendant's argument works as follows. By R3 and F2 and DMP, the conclusion plaintiff was informed ¬stairway in a safe condition follows. By this conclusion and R4, the conclusion that the owner is not liable follows.

The ultimate conclusion of the plaintiff's argument is the opposite (strong negation) of the ultimate conclusion of the defendant's argument. Each side has a valid chain of reasoning leading to its conclusion based on DMP and on a factual proposition thought to be provable by the that side. So far the logic of the reasoning seems to be that of the standard legal case in civil law. The problem is to make a determination of which parts of the reasoning on each side can be designated as presumptions.

It might appear that on the Prakken and Sartor theory, R3 is the presumption. That would not seem to be correct, because R3 is about this particular instance in which the defendant allegedly mailed a certain letter to the plaintiff. On the Prakken and Sartor theory, the presumption is the general rule to the effect that if a letter was properly addressed, stamped and deposited in an appropriate receptacle it will be presumed that it has been received. This is a general rule of law, as opposed to a particular statement in a given case that is an instance of the application of this rule. Thus, it would seem

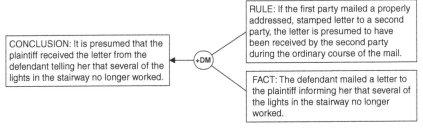

FIGURE 3.2. The Mailed Letter Rule in an Inference in the Stairway Case

correct to say that the presumption in the case, as modeled by the Prakken and Sartor theory, does not appear specifically in the set of facts and rules stated above. Rather it is a general legal rule that backs or supports R3.

However, it seems more natural to say that the presumption in the case is the statement that the plaintiff was informed that the stairway was not in a safe condition. In the chain of reasoning on the defendant's side, this statement has the role of an interim conclusion generated by R3 and F2 using DMP. This view of the argumentation accords with the view that the presumption is to be identified with the statement that is derived by inference from a given fact of a case along with a general defeasible rule that applies to the fact. The given fact is the premise of the presumptive inference. We can describe the situation by saying that a general defeasible rule applies to this fact, thereby creating a presumption. The presumption is created by the rule of inference and the given fact by applying the rule to the fact, generating the presumption as a conclusion of the inference.

The most natural way of representing how the presumption is generated in this case is through an examination of its basic inferential structure, as shown in Figure 3.2.

That this is the more natural way to describe the reasoning in the dark stairway case, and probably many other cases of presumptive reasoning as well, does not really detract, at least very much from the Prakken and Sartor theory, however. The reason is that this theory is supposed to be a formal model of presumption that is based on nonmonotonic logic and meant to be used for purposes of artificial intelligence as applied to legal reasoning. Because it is a formal model that is meant to be implemented in computing systems, it need not represent the most natural way of speaking about presumptions in everyday conversational argumentation, or even for that matter in legal argumentation of the kind found in a trial in civil or criminal law.

Finally, we need to comment on how the presumption in the case is related to the burden of proof. It is a civil case, so the burden of proof is distributed equally between the two parties. The standard of proof is "preponderance of the evidence," meaning that whichever side has the stronger argument wins. Once the plaintiff has put forward her argument, provided her rules and facts are supported by sufficiently strong evidence, she will win the trial, unless the defendant can put forward a stronger argument. At the point where her argument above is offered, we are still in the argumentation stage. The outcome is not decided yet, but unless the defendant puts forward a stronger counterargument, he loses. What this means is that the plaintiff's argument has placed an evidential burden on the defendant's side, as soon as she puts her argument forward. His counterargument, assuming it is strong enough, meets this evidential burden. The conclusion of the plaintiff's argument is `owner liable`. The support of this argument by the plaintiff by admissible evidence creates an evidential burden on the side of the defendant in a way that is typical of evidential legal argumentation in a trial.

To meet this evidential burden, the defendant uses the device of presumption. The defendant's problem is that he lacks evidence to prove the proposition `plaintiff was informed ¬stairway in a safe condition`. All he can prove is the proposition `defendant mailed letter to plaintiff, saying lights in the stairway don't work`. What use is this? It would be of no use as evidence, except for the existence of the defeasible legal rule R3, `defendant mailed letter to plaintiff, saying lights in the stairway don't work → plaintiff was informed ¬stairway in a safe condition`. This rule can be applied to the fact that such a letter was mailed, giving rise by defeasible inference to the presumption that the plaintiff was informed that the stairway was not in a safe condition. This evidence supports the conclusion `¬owner liable`. Even though the reasoning is merely presumptive, it may represent strong enough evidence to meet the evidential burden set into place by the plaintiff's argument, shifting the evidential burden of proof back to her side. This shifting back and forth of such an evidential burden could take place many more times during the trial.

7. Combining the Inferential and Dialogical Components

There have been differing views in philosophy on how the term 'inference' should be defined. In logic textbooks, this term is used commonly to represent a structure made up of a set of propositions one of which is designated as the conclusion. The other propositions are called the 'premises of the inference.' The conclusion is said to be drawn from the premises, and the premises constitute a set of assumptions. Abstractly, the inference may be defined as a relation on the set of propositions making up the premises and conclusion. A typical inference is a sequence of the following sort: all men are mortal; Socrates is a man; therefore Socrates is mortal. This inference is classified as a deductive one, as contrasted with inductive inferences. Inductive inferences are nonconclusive and nonmonotonic, meaning that if new premises are added, the inference may fail to hold.

The third category of inference that is recognized sometimes is also nonmonotonic. A typical inference of this sort is a sequence of this kind: birds fly; Tweety is a bird; therefore Tweety flies. This type of inference is sometimes called defeasible, meaning that it is subject to failure in a case of an exception that could not be predicted in advance. For example, if Tweety has a broken wing, the two premises may be true, but inferring the conclusion from them no longer holds. This meaning of the term "inference" could be called the logical interpretation.

According to a contrasting interpretation (Brown, 1955, 360), an inference can never, strictly speaking, be deductive. On this interpretation, inferences do not have premises (358). Instead, inferences should be described as arising from facts or supposed facts (358). An example of an inference is the following: "Riding into town, he saw most of the flags at half-mast and inferred that some well-known person had died" (Brown, 1955, 355). This inference is different from a deductive argument, or for that matter, from an inductive argument, and can be identified with what is now called defeasible inference in the literature on AI, like the Tweety example. If it is known that it is a general practice in the area to only put the flags at half-mast when some well-known person has died, the inference is a reasonable one. However, it could default in some circumstances. For example, suppose that a recent practice is sometimes in place of putting the flags at half-mast when any soldier has died in a foreign engagement on that day. The soldier would not normally be a well-known person; in such an instance the inference would default.

Whinery (2001, 2) noted that the courts on occasion have used the terms 'inference' and 'presumption' synonymously. He drew the distinction in terms of the notion of probative value. The probative value of an inference, or a proposition that is a premise in an inference, may be defined as its capability to increase probative weight as evidence of a proposition that is the conclusion of an inference. A presumption created by a rule of law can have probative value, but the way that this probative value arises in the case of a presumption is always different, according to Whinery, from the way it arises in the case of an inference not based on a presumption. In the case of an inference not based on a presumption, the probative value of the conclusion drawn arises only from the probative force of the evidence. However, in the case of a presumption it arises from a rule of law. On this analysis, a presumption is a kind of inference, or at least is based on a kind of inference, but it is a special kind of inference in which one premise is a rule of law. A rule of law may carry probative weight, but it does so in a way that is different from a factual generalization. A factual generalization carries probative weight because it is supported by evidence that can be brought forward by one side and questioned by the other side.

According to Ashford and Risinger (1969, 165), there are some judicial limits that have been set on the use of presumptive inferences in law following due process requirements of the fifth and fourteenth amendments that "void the operation of presumptive language which works in an unreasonable or capricious manner." They cite two cases showing that presumptive language may not be used to circumvent constitutional rights: *Mobile J. & K.C.R.R. v. Turnipseed* (219 U.S. 35 (1910)), and *Bailey v. Alabama* (219 U.S. 219 (1911)). In another case, *Leary v. United States* (395 U.S. 6, 36 (1969)), a court proposed a constitutional test called the "rational connection test" that is supposed to supply a more stringent standard to presumptive reasoning. A widely known statement of this test comes from yet another case, that of *Tot v. United States* (3 U.S. 463, 467 (1943)):

Under our decisions, a statutory presumption cannot be sustained if there be no rational connection between the fact proved and the ultimate fact presumed, if the inference of the one from the proof of the other is arbitrary because of lack of connection.

According to the requirements of the dialogical theory, the rational connection between the fact proved and the ultimate fact presumed must have five inferential components. The first is a form of argument or argumentation scheme. For example *modus ponens* is a form of argument. In the examples treated here, the form of argument is that of defeasible *modus ponens*. The second component is a general rule that is meant to fit a particular case. One premise is a generalization (general rule) that takes the form of a general statement in the form of a conditional. The rule is often of a practical sort, which stems from efficient means needed to move a discussion or inquiry forward. In the examples treated here, it is a defeasible generalization. Some argue that it is always a defeasible generalization because there are no conclusive or irrebuttable presumptions. However, this issue is controversial. The third component is a set of factual premises describing the particulars of a given case at issue. The fourth component is the proposition derived from the first three components by defeasible reasoning. It is often called the presumption that arises in the particular case. The fifth component is that of the probative function of an inference. Each proposition in an inference, the premises and the conclusion, may be said to have probative weight of the kind that can be ranked in an ordering. In other words, some propositions have equal probative weight, some propositions have more probative weight than others, and some have less probative weight than others. The probative function of an inference refers to the use of an inference with premises that carry probative weight to increase the probative weight of the conclusion (Walton, 2002, 214–216). A presumption must have all five components to be a rational connection.

As indicated earlier, presumption also has dialogical components. The first is that there is some sort of ongoing discussion or investigation underway, and there are various participants taking part in it. In the simplest kind of case we study here there are only two participants, called the proponent and the respondent, and the issue to be resolved by the discussion is a conflict of opinions between them. The second component is the burden of persuasion. Each side has a thesis, a particular proposition that this side is required to prove to some standard proof in order to win the discussion. The third component is that the dialogue is composed of three stages. Matters of which side has to prove what in order to win are set at the opening stage of the dialogue. The fourth component is the burden of persuasion. At the opening stage, it is made clear that each side has a burden of persuasion. The burden of persuasion is composed of three elements stating (i) a thesis, (ii) which side has to prove that thesis and (iii) what standard of proof has to be met. When a side puts forward arguments of the kind that successfully fulfill its burden of persuasion, that side is the winner in the dialogue. The question of which side won or lost, by fulfilling its burden of persuasion or not, is cited at the closing stage of the dialogue. The middle stage, between the opening and closing stages, is called the argumentation stage.

The relationship between the inferential and the dialectical components of presumption can only be properly appreciated once one fits the inferential component into the setting of the dialectical component. You might think that the notion of a presumption is completely defined by the five inferential components, but actually this is not so, because there are plenty of defeasible inferences from a fact and rule to a conclusion about a particular case derived from the fact and rule as premises. This distinction has been observed in law (Allen and Callen, 2003). The difference between a presumptive inference and an ordinary defeasible inference that is not presumptive in nature relates to the notions of burden of proof and standard of proof, factors that vary in different contexts of dialogue. In a trial, for example, there are two sides. The claim of the one side is opposed to the claim made by the other. When one side puts forward an argument that has probative value, that argument supports its claim, and thereby rebuts the claim made by the opposed side. When this happens during the argumentation in a trial, it is often said that the burden of proof has shifted to the opposing side. Bringing forward a presumption can have the same effect. Thus, the effect of bringing forward an inference based on a presumption, if the conclusion of the inference carries evidential weight, is that it places what might be called an evidential onus or burden on the opposing side. This phenomenon is the link between presumption and burden of proof.

A presumption may be defined as a plausible inference based on a fact and a rule as premises, where the premises are insufficient to support the conclusion in accord with the link or warrant presenting the argumentation scheme joining the premises to the conclusion, and where a further boost is needed to gain a proper acceptance of the conclusion. But what is meant by the term 'insufficient' in this definition? What is meant is that the argument fails to meet the burden and standard of proof that should be required to make it sufficient to prove the conclusion. But what does this mean? When is an argument sufficient to prove its conclusion? The answer to this question can only be sought in the notions of burden of proof and standard of proof. These notions vary with the type of dialogue in which the argument occurs and the stage of the dialogue at which the argument was put forward (Krabbe, 2003). Hence, dialectical notions are required to fully define the concept of a presumption.

The need for making a presumption arises during the argumentation stage when a particular argument is put forward by one side. Typically what happens is that a problem arises because there is some particular proposition that needs to be accepted at least tentatively before the argumentation can move ahead. But at that point in the dialogue, this proposition cannot be proved by the evidence that is available so far. That evidence is insufficient, and the circumstances are such that collecting it would mean a disruption of the dialogue, for example, because it might be very costly or take too much time to conduct an investigation to prove or disprove this proposition by the standards required for properly accepting or rejecting it. It is precisely in this kind of case where the notion of presumption comes in. The proposition can be tentatively accepted as having the status of presumption even though the evidence supporting it at that present point in the dialogue is insufficient for accepting it. The reason for accepting it, typically a practical one, is to be found in the rule premise and the factual premise that support it as a conclusion.

Hence, it is very important to recognize that the notion of sufficiency for acceptance is an essential component in the definition of presumption. The presumption is not just a general rule that can be applied to a particular case. Nor is it just the application of such a rule to a factual proposition to draw an inference leading to a factual assumption. It is the use of this kind of inferential setup for a particular purpose in a dialogue, in a case where the factual assumption cannot be proved, and therefore accepted, by this standard of proof appropriate for the argument at the stage at which it occurs in the dialogue. It is important to recognize that a presumption is not a conclusion that has been proved. In contrast, it is an assumption that cannot be proved, as the evidence has been collected at this point in an investigation, but that warrants acceptance anyway.

8. Application of the Dialogical Theory to Examples

The key problem for the new theory is to define presumption in such a way that enables a distinction to be drawn between presumption and any other kind of defeasible *modus ponens* inference. Defeasible *modus ponens* inferences are extremely common in legal reasoning and in conversational argumentation generally, whereas presumptions are comparatively rare. On the Ullman-Margalit model, a presumption is defined as an inference based on two premises, one of which is a fact and the other of which is a generalization. The new theory finds this model acceptable as far as it goes, but the problem is that a deeper theory is needed in order to distinguish between the most common kinds of inferences based on facts and rules and those that can be distinguished as distinctively presumptive inferences.

The new dialectical theory solves this problem by setting out eight requirements for an inference to be a presumption.

1. It must be a linked argument of the defeasible *modus ponens* form.
2. One premise, called the fact premise, states some fact concerning a particular case at issue.
3. The other premise, called the rule premise, states a general rule that has a conditional form.
4. The context is that of a persuasion dialogue in which a burden of persuasion has been set at the opening stage.
5. The conclusion drawn from the fact and the rule is put forward to gain the respondent's acceptance of it.
6. The argument is not sufficiently strong, based only on the evidence supporting the two premises to shift a burden of production to the respondent's side.
7. The presumptive rule has a practical justification in line with the goal of the persuasion dialogue.
8. The argument is sufficiently strong, with the practical justification counted in, to shift a burden of production to the respondent's side.

The first five requirements are not enough by themselves to distinguish between presumptions and many of the most common kinds of inferences found both in legal reasoning and everyday conversational argumentation. It is the last three requirements that mark off presumption as representing a distinctive kind of reasoning that can be brought to bear when the evidence is not strong enough in a given case as a basis for resolving a disputed point needed to move the argumentation ahead so that the ultimate conflict in the dialogue can be resolved.

The first three requirements are inferential in nature. They describe the characteristics of the inference, that it must be a linked argument, that it must have certain kinds of premises and that the conclusion is drawn from the premises by a defeasible inference of a certain sort. The remaining five requirements are contextual in nature. They pertain to the context of dialogue in which the inference is used to move a chain of argumentation forward toward its goal in a dialogue. The dialogue structure provides a conversational setting in which the inference is used as a species of argumentation.

Now let's apply this theory to the case in which a person is declared dead by law in order to distribute his estate, based on the fact that he has disappeared for five years and the legal rule that if a person has disappeared for a period of five years or greater, he or she can be declared dead for legal purposes. The claim made in such a case, let's say, is that the proceeds of the estate should be divided up and handed out according to what the person stated in his will. One of the elements of this claim is that the person has to be dead. This is a civil case in which the claim has to be proved by the proof standard called preponderance of the evidence. To fulfill his burden of persuasion in the case, the claimant has to produce a strong enough argument to meet the standard of proof. If his argument meets this requirement, he has filled his burden of production, and the burden of production shifts to the other side. A burden of production now shifts to the other side to produce some evidence that person is not dead. The presumer[3] side has fulfilled its burden of production, and if the presumee side does not fulfill its burden of production, and the presumer proves all the other elements to the required standard, the presumer fulfills its burden of persuasion. The bottom line is that the party who claims that the proceeds of the state should be divided up and handed out according to what that person stated in his will has won the case.

Now let's apply the theory to the stairway case. The plaintiff sued the building owner, claiming he did not keep the stairway in a safe condition. In civil law, this claim needs to be proved using the preponderance of the evidence standard. Her ultimate *probandum* was the proposition that he did not keep the stairway in a safe condition. Hence, the burden of persuasion was on her to prove that the stairway was unsafe. Now we need to look at the arguments on both sides. The argument the plaintiff gave to prove her claim that the stairway was unsafe was that the lighting did not work properly. This argument would be defeated, however, if the defendant had informed her beforehand that some of the lights in the stairway no longer worked. To defeat it, the defendant argued that he had sent the plaintiff a letter informing her that several of the lights in the stairway no longer worked. The inferential structure of the plaintiff's argument is shown in Figure 3.3. DM represents the scheme for defeasible modus ponens.

The evidential reasoning in the plaintiff's argument is straightforward. The two premises in the top on the right are used to derive the conclusion that the stairway was unsafe by a *modus ponens* inference. At the next level in the chain of reasoning, three additional premises are used to support the

---

[3] Using the terms plaintiff and defendant can sometimes cause the reader to lose track when the burden of production is shifting back and forth from one side to another repeatedly over several moves in a dialogue. For this reason we have invented the artificial terms 'presumer' and 'presumee' to name each side.

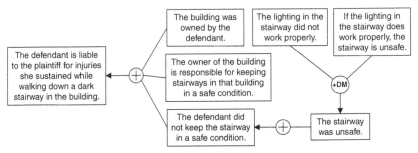

FIGURE 3.3. Inferential Structure of the Plaintiff's Argument

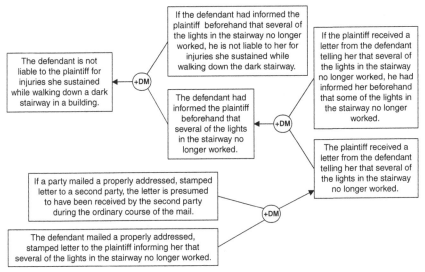

FIGURE 3.4. Inferential Structure of the Defendant's Argument

plaintiff's argument that the defendant is liable to the plaintiff for injuries sustained while walking down the dark stairway. The reasoning is based on evidence and no presumptions are required. This could be called a case of normal evidential reasoning, meaning that no presumptions are needed to boost up weak points in the chain of reasoning needed to establish the conclusion.

In contrast, in the defendant's argument, a presumptive inference is needed to establish the conclusion. The inferential structure of the defendant's argument is shown in Figure 3.4, displaying that the presumptive inference fits in with the other inferences in the chain of reasoning.

In the defendant's argument, as shown in Figure 3.4, there are three steps of inference. The inference from the first two premises shown at the bottom left to a conclusion that the plaintiff received a letter from the defendant telling her that several of the lights in the stairway no longer worked is a presumptive inference. But this presumptive inference is combined with the two normal *modus ponens* inferences in the argument diagram. Normal evidential reasoning is not enough here. To fill the gap, a presumptive inference has to be used along with the two other inferences.

The conjecture can be put forward here that the Prakken-Sartor theory, an account of legal presumption meant to be suitable for artificial intelligence, can be combined with the dialogical theory, incorporating the Prakken-Sartor theory as its logical component to display the connections between burden of persuasion, evidential burden and presumption.

There is an additional component that needs to be added to the dialogical theory of presumption. What needs to be added is the requirement that the inference is sufficiently strong to satisfy the evidential burden of the party who put it forward as a presumption (the proponent). In other words, to be a presumption, it must shift an evidential burden from the proponent's side of the dialogue to the opponent's. The effect in the dialogue must be that the party doubting or denying the presumption that has been inferred (the opponent) must give some argument against it, or else he will risk losing the exchange, leaving it so that the proponent's argument appears more plausible than his. To sum up then, on this theory, the notion of a presumption may be defined as follows. A presumption is (a) a proposition put forward by a proponent in a dialogue indirectly by drawing it (or allowing the respondent to draw it) as a conclusion from a factual proposition and a general rule, (b) on the basis of a defeasible inference,[4] (c) to gain the respondent's acceptance of it, (d) of sufficient strength so that it meets an evidential burden that shifts the burden of proof in the dialogue from the proponent to the respondent.

There is much discussion in the legal literature as well as the literature on argumentation theory concerning the distinction between rebuttable and nonrebuttable presumptions, sometimes also drawn as a distinction between permissible inferences and conclusive presumptions (Allen and Callen, 2003, 936). For reasons of space, we do not include a discussion of this distinction, and are compelled to leave it as a subject for further research. Because of the importance of the subject, and the abundant literature on it, we cannot deal with it in the scope of this chapter, even though the theory put forward in the chapter has significant implications for rethinking this distinction.

---

[4] Although presumptions of the most common sort are generally defeasible, exceptions need to be made for conclusive presumptions in law. Courts have presumed that if a child is under seven years of age, she could not have committed a felony (Strong, 1992, 451). We leave open the issue of whether so-called conclusive presumptions of this sort require an extension or modification of the theory.

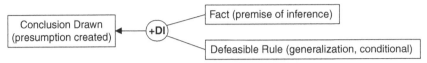

FIGURE 3.5. Inferential Structure of Presumptive Reasoning

This theory is meant to model presumptive reasoning in everyday conversational argumentation, but derives its inspiration from how burden of proof and presumption work as legal concepts in the context of the trial. Ordinary conversational argumentation is taken normally to be based on a dialectical framework in which two parties each criticize the arguments of the other. In contrast, legal argumentation in a trial is clearly a three party dialogue structure in which there are two opposed sides along with a third party trier who decides the outcome of the dialogue. How presumption works in a trial needs to be analyzed in relation to the interplay among these parties. How presumption operates in this kind of structure can be illustrated by McCormick's account (Strong, 1992, 460) of the effects of presumption in a civil jury trial. It can happen either where one party or the other moves for a directed verdict, or when the time comes to instruct the jury. The defendant's motion for a directed verdict will be denied by the judge even though the plaintiff has failed to offer any evidence to support its claim, if the plaintiff can show that the basic facts give rise to a presumption of that claim: "the jury will be instructed that if they find the existence of the basic facts, they must also find the presumed fact" (461). We can see how presumptive reasoning is used in the context of the trial as a way of finding a substitute for meeting an evidential burden in some cases, as shown in the inferential structure of the defendant's argument in Figure 3.5.

9. Conclusions

On the new dialogical theory, presumptive reasoning is defined as a special kind of inferential structure in which there is a premise, a conclusion and an inference leading from the premise to the conclusion. On this theory, presumptive inference is generally taken to be defeasible. However, room is left for the possibility that there can be so-called necessary presumptions based on deductive reasoning, a kind of reasoning that is not defeasible. In presumptive reasoning, there is a general premise typically taking the form of a conditional statement or generalization, called a rule in AI, and the rule is applied to a so-called fact. The fact is a particular statement that is taken to be true in a given case. The rule is applied to the fact generating a conclusion by a defeasible form of inference. The application of the rule to the fact gives rise to another particular statement called the presumption. The basic inferential structure of presumptive reasoning is simply displayed in Figure 3.5, where DI can represent any defeasible argumentation scheme.

It is important to recognize that the two premises at the right go together to support the conclusion, and so the argument structure is that of a linked argument. In contrast, in a convergent argument, each premise represents an independent support for the conclusion. In a linked argument, the premises function together.

According to Prakken and Sartor (2006), presumption is to be identified with the defeasible rule operative in the inferential structure shown in Figure 3.5. On the dialogical theory, in contrast, the presumption is to be identified with the conclusion of the inference. The decision between the two theories is a choice of which language to adopt. Perhaps one choice is better for using artificial intelligence to model legal reasoning while the other is better to build an argumentation theory for reasoning and everyday conversational argumentation. Either choice is better than leaving the notion of presumption in its current vague and slippery state. Whatever choice is made, the most important thing is to recognize that presumptive reasoning has this general structure.

A presumption is a very common device used to assist a dialogue to move forward toward reaching its goal by argumentation used for that purpose. Usually the requirement of burden of proof demands that for every claim made or argument put forward reasons have to be given to back up that claim or argument in case it might be questioned. Backing up every argument with convincing reasons to support it could be extremely costly in time and effort in many instances, even so burdensome that it might delay that dialogue from moving forward or block it entirely from continuing any further. The function of a presumption, according to the dialogical theory, is to remove this potential blockage and enable the dialogue to move forward from a given point during the argumentation stage. In practical terms, the function of presumption is to save time and money and effort in communication. Any kind of communication is only possible, according to Grice (1975), if both parties contribute in a collaborative manner to moving the dialogue along toward its goal by making the right kinds of moves at the right point in the dialogue with the kind of moves needed. This principle is the fundamental rule or chief conversational postulate in any Gricean style theory of collaborative conversation suitable to frame the dialogical theory of presumption.

Using the two examples of the presumption of death and the letter delivered in the mail, it has been revealed in this chapter how the two concepts of burden of proof and presumption can precisely be distinguished. Burden of proof is a global factor that pertains to a whole dialogue, and can affect all four stages. Presumption is a speech act that fits in between assertion and assumption, and other speech acts that are used by both parties to a dialogue at the argumentation stage. It is vitally important to draw

TABLE 3.1. *Properties of Speech Acts Relating to Presumption*

Speech Act	Goal	Precondition	Definition	Burden	Effect
Assertion	To put forward a claimed proposition for assent	Can be made at any appropriate point	Commits speaker to holding the proposition as true	Must justify (give an argument in support) or retract	Proposition added to speaker's commitment set
Assumption	To see where it leads by a chain of argumentation	Can be made at any appropriate point	No need for commitment to truth or even plausibility	No burden of proof	Respondent must accept it if it is relevant
Presumption	To secure tentative assent	Practical need to move forward in absence of proof	Commits speaker to holding a proposition as not false	Negative burden of showing the proposition is not false	Respondent must accept for now unless he can show falsity
Argument	To gain commitment by offering reason to accept	Doubt by the other party that can be addressed by offering reasons	Offering premises in support of the truth of the conclusion	Must defend the argument if attacked, or retract it	Helps move speaker's side forward to meeting burden

116

a three-way distinction among three of these speech acts, those of assertion, assumption and presumption and contrast them with the speech act of putting forward an argument. When you make an assertion, you are obliged to offer justification for it, or retract it. You are free to make an assumption at any time, even if it can't be proved, or even if you and others know it is false. When you make a presumption, you are not obliged to offer a justification for it, but you are obliged to give it up if the other party can disprove it.

Those familiar with fallacies will immediately see the connection with the *argumentum ad ignorantiam*. You presume that something is true without having to prove that it is true, but you do have to retract it if the other party can prove it is false. Table 3.1 summarizes the key properties of these speech acts. As opposed to a positive burden to show a proposition is true there is a negative burden of showing the proposition is not false.

More will be explained in the next chapter about the importance of these speech acts as fundamental units necessary for building formal models of dialogue for argumentation. For the present, we only need to see how presumption is different from assumption, assertion and argument (especially of the defeasible sort).

In the presumption of death example, the court cannot prove that the missing person is dead, but has some negative evidence to support that proposition, and has no positive evidence that proves it is false. If that proposition were to be asserted, or to be argued for in an attempt to prove it, either of these speech acts could easily fail if challenged. However, the proposition can be presumed as a tentative basis for action, until such time as new evidence comes in showing that it is false. In the letter example, the sender may not be able to prove that the addressee received it, but unless he can prove that he did not receive it, she can put forward a presumption that he did, as part of her argument in the case, and the court will tentatively accept it.

We can see in such cases how presumption is so closely connected to burden of proof. The presumption shifts a burden to the other side to disprove it, or the proposition becomes lodged into place as a commitment of both sides. This shifting takes place at a local level, and it can be reversed at a later point in the argumentation stage. But it puts in place a mechanism for tentative acceptance of the proposition to help the dialogue move ahead towards its goal. This property is typical of presumptions. We use them to help a dialogue move ahead toward resolving a conflict of opinions on a burden of proof basis, even though we lack knowledge of a kind that would provide us with certainty, or with enough evidence to resolve the issue by proving one side's claim true or the other's false.

According to the theory of the speech act of presumption in (Walton, 1992, 278–282), summarized in Chapter 1, Section 4, there are four conditions governing this speech act in a dialogue. A is the proposition that is contained in the presumption. The preparatory condition states that a proponent and a respondent are engaged in a dialogue in which A is a relevant assumption. The placement condition states that A is brought forward for acceptance by the proponent who made the presumption, that the respondent has an opportunity to reject A, and that until he rejects it, A becomes a commitment of both parties. The retraction condition permits the respondent to retract commitment to A at any point in the following dialogue, provided that he can give evidence to support such a rejection. The burden condition states that at any given point in the dialogue, the proponent has the burden of showing that assuming A has some practical value in moving the discussion forward, and that the respondent must let the presumption stay in place as long as it is useful.

10. Directions for Future Research

The dialogical theory of presumption presented in this chapter has raised a number of problems that cannot be solved within the scope of a single chapter or article. Placed in a context of the recent literature on presumption and burden of proof in artificial intelligence, alongside the other theories of presumption that have been put forward in the field of argumentation studies, the dialogical theory offers resources for approaching these problems in a different way.

One problem we have not addressed in this chapter is how to rebut a presumption in legal argumentation. This is an important problem to be addressed in future work. It has not been addressed in this chapter because there are theoretical problems about how to define the notion of a rebuttal that are significant enough to merit a separate investigation. Two theories already exist in the legal literature (Park, Leonard and Goldberg, 1998, 109–111). According to the bursting bubble (Thayer-Wigmore) theory of presumption, presumptions are "like bats flitting in the twilight, but disappearing in the sunshine of actual facts" (109). It says that a presumption should have no effect once "rebutted" with evidence challenging the presumed fact. In the letter example, suppose that the plaintiff did not challenge the defendant's proper addressing, stamping and mailing of the letter, but testified that during the whole period, she picked up and diligently read her mail each day, and she never saw the letter. On the bursting bubble theory, the presumption that the plaintiff received the letter is cancelled. The jury would now be left "to apply its sense of logic and experience" to determine whether the plaintiff received the letter or not (110). According to the Morgan-McCormick theory, once a presumption is raised by its proponent, the burden of proof shifts to the opponent (111–112), or otherwise the presumption stands. This theory holds that the bursting bubble theory gives too "slight and evanescent" an effect to presumptions (111). On this theory, if the jury finds that the defendant properly addressed, stamped, and mailed the letter sufficiently in advance of the accident, the plaintiff must prove it more likely that she did not receive the letter or she must suffer a finding that she did (112). Which theory is right depends on how the notion of rebuttal (also often called refutation, attack, argument defeat, and so forth) is to be defined. It also depends on general issues in argumentation theory on how arguments are to be evaluated.

Another general problem that has been posed in this paper is how the notion of presumption relates to argumentation schemes. The Prakken-Sartor theory can be combined with the dialogical theory to solve the problem of the relationship between presumptions and argumentation schemes. To give an example, consider the following reformulation of the argumentation scheme for argument from expert opinion (Reed and Walton, 2003).[5] In this version, a conditional premise that links the major to the minor premise has been added.

Argument from Expert Opinion

> Major Premise: Source *E* is an expert in field *F* containing proposition *A*.
> Minor Premise: *E* asserts that proposition *A* (in field *F*) is true (false).
> Conditional Premise: If source *E* is an expert in a field *F* containing proposition *A* and *E* asserts that proposition *A* is true (false), then *A* may plausibly be taken to be true (false).
> Conclusion: *A* may plausibly be taken to be true (false).

In this version, the additional premise was called a conditional premise in (Reed and Walton, 2003), because it takes the form of what is called a conditional proposition in logic.

Let's apply this version of the scheme to an example. Let's say that Jason is a forensic expert in the field of ballistics evidence and that he has testified that the bullet found in the victim's body matches the defendant's gun. Let's say that these two propositions are accepted as factual in a particular case. Given these two facts, it may be taken as a presumption that the bullet found in the victim's body matches the defendant's gun. The presumption, as commonly said, arises from these two facts. But what is the logical structure whereby the two facts give rise to the presumption by some sort of identifiable logical inference? According to the dialogical theory, the structure is the argumentation scheme for appeal to expert opinion. If we look at the scheme, we can see that the conditional premise of the scheme acts as the rule in the Prakken-Sartor theory. The defeasible rule, shown to have the conditional form in version 2 of the scheme for appeal to expert opinion, enables the conclusion of the scheme to be drawn by a defeasible *modus ponens*. Thus, the presumption is raised that the bullet found in the victim's

---

[5] The version in (Reed and Walton, 2003) uses a variable *S* for the subject domain of the proposition.

body matches the defendant's gun. This presumption would then shift an evidential burden in the context of a murder trial in which the expert ballistics testimony is evidence.

The structure of the inference in the example can be modeled in the Prakken-Sartor theory as the following sequence of reasoning.

```
F1: Jason expert
F2: Jason testified bullet matches weapon
R1: Jason expert & Jason testified bullet matches gun →
    bullet matches gun
```

Applying defeasible *modus ponens* to F1, F2 and R1 yields the conclusion `bullet matches gun`. Through this simple example we can see how the presumption arises in the case through the application of the Prakken-Sartor theory and the dialogical theory combined. We can now see how presumptive argumentation schemes, like the one for appeal to expert opinion described in Chapter 1, Section 6, can justifiably be classified under the category of presumptive reasoning. Such schemes can generally be so classified because the conditional premise, the generalization implicit in the scheme, functions as a defeasible rule of the kind specified in the Prakken-Sartor theory.

One of the most important features of the dialogical theory is that it brings out the relationship between presumption and evidence. The burden of persuasion set at the opening stage of a dialogue implies that the general default rule applies through the argumentation stage of the dialogue – the party who asserts, or makes a claim, must back it up with evidence. Putting forward a presumption, as opposed to making a claim in the form of an assertion, is an exception to this general rule. According to the dialogical theory, a presumption can be set in place as the conclusion of an implicit argument based on a factual premise and another premise that is a default rule. Thus, the speech act of putting forward a presumption during the argumentation stage has a structure that is very similar to the speech act of putting forward an argument. This structure is brought out very well by the new dialogical theory. Thus, the theory displays the structure whereby presumption has a function of presenting evidence comparable to the way evidence is presented in law, namely by providing an argument that gives reasons to back up a disputed claim. Presumption can be seen as a kind of argumentation device that provides a reason for tentatively accepting a claim in the absence of evidence to the contrary.

Prakken and Sartor, following Williams, call the burden of proof set at the opening stage of a legal dialogue, like that of a trial, the burden of persuasion. The question arises whether burden of persuasion only applies in persuasion dialogue, or whether it applies in other types of dialogue as well, like deliberation, negotiation, inquiry, information seeking and eristic dialogue. It would seem to be a likely hypothesis that it does not apply in some of these other types of dialogue. For example, in negotiation, each party is trying to get the most of what it wants, and so there would seem to be no place for matters of burden of proof to arise, except where the dialogue shifts to a persuasion interval. If this general hypothesis is right, it follows that burden of persuasion is a unique characteristic of persuasion dialogue, as contrasted with these other types of dialogue. If this is so, it may help us distinguish between persuasion dialogue and, say, deliberation dialogue.

On the Prakken-Sartor approach, burden of production and tactical burden of proof only arise where there is a burden of persuasion in a dialogue. If this hypothesis is right, then it seems likely also to be true that these two concepts have no place in types of dialogue other than persuasion dialogue. However, it would appear that burden of proof has a recognized place in at least one type of dialogue, namely the inquiry (Walton, 1996). So it may well be that something like burden of proof, which should not be called burden of persuasion, plays a role in the other types of dialogue as well. How notions comparable to burden of persuasion work in these other types of dialogue is a question taken up in Chapter 7.

4

Shifting the Burden of Proof in Witness Testimony

In the previous chapters we have discussed how to represent the operation of critical questions in a formal and computational model that can incorporate argumentation schemes as well as their accompanying critical questions. In order to illustrate how this works the example of the scheme for argument from expert opinion has been used. The problem is to classify the critical questions as assumptions or exceptions in order to properly reflect the distribution of the burden of proof between the party who put forward the argument and the other party, the respondent who is raising critical questions about the argument. Is this problem merely a technical problem of how to model argumentation by the use of defeasible argumentation schemes? Or is it a problem that could arise in a real case of argumentation? In Chapter 4, a legal case concerning how to logically represent critical questions appropriate for argument from witness testimony is studied that illustrates the problem of how to arrive at a decision to properly assign a burden of proof to the one side or the other.

In this case, the Oregon Supreme Court overturned the previous procedures for determining the admissibility of eyewitness identification evidence. The decision to change the law was based on recent research in the social sciences concerning the reliability of eyewitness identification, and by considerations put to the court by the Innocence Network, an organization dedicated to the study of unjust convictions. In some cases it can be quite difficult for the courts to make a decision on burden of proof, and in some of these cases a ruling is made that can act as a precedent when the same kind of decision about burden of proof arises in a comparable case. In Chapter 4, a more challenging kind of case is studied in which a change was made in the normal way of dealing with burden of proof in criminal trials. This change was prompted by a gradually growing body of scientific evidence suggesting that witness testimony evidence is much more fallible in certain respects than was previously thought.

Scientific research over the past thirty plus years has identified many kinds of bias in witness testimony evidence arising from suggestions put in place by exposure to misleading information. This recognition of the fallibility of witness testimony as a form of evidence has led to changes put in place by the Supreme Court of Oregon shifting the burden of proof from the defendant to the state (the prosecutor) to prove that witness testimony is admissible. This chapter uses the Carneades Argumentation System (Gordon, 2010) to model the evidential structure of the two cases where the changes were made, and to study the implications of them for formulating the argumentation scheme for argument witness testimony for use in argument visualization tools.

The first section reviews the relevant features of the Carneades Argumentation System and shows how it uses the scheme for argument from witness testimony to model the bias critical question. The second section uses Carneades to analyze and evaluate the central structure of the evidence in the first of the two cases studied. The third section offers a brief overview of biases found in social science research to show factors in the fallibility of witness testimony as evidence. The fourth section gives a brief overview of the structure of the evidence in the second case study, so that by comparing it with the first case the reader can appreciate the rationale behind the changes made by the Oregon Supreme Court. The fifth section shows how the ruling made in the second case shifted the burden of proof from the defendant to the prosecution to establish admissibility of eyewitness evidence. In the sixth section specific recommendations are made on how to reconfigure the argumentation scheme for argument from witness testimony. The seventh section shows how the witness testimony scheme is based on prior schemes for perception and memory. Section 8 shows how burden of proof for admissibility fits into the dialogue structure of a trial, with opening and closing stages. Section 9 connects admissibility with the general notion of relevance in argumentation theory by showing how argumentation takes place at different stages of the trial procedure. Section 10 provides conclusions to the chapter.

1. Witness Testimony in the Carneades Argumentation System

In the previous chapters it was shown how argument mapping and argumentation schemes are fundamental tools that need to be combined for modeling argumentation in artificial intelligence and law. How the scheme for argument from witness testimony is applied in some argument mapping systems is a case in point. In the catalogue of argumentation schemes in the argument visualization tool Rationale,[1] three argumentation schemes attributable to Pollock (1995) are given in a simple format in which each

---

[1] http://rationale.austhink.com.

scheme has only one premise. The witness testimony scheme has the premise "Witness *W* says that proposition *A* is true," and the conclusion *A*. The perception scheme has the premise "Having a percept with content *A*" and the conclusion *A*. The memory scheme has the premise "Recalling *A*" and the conclusion *A*. Each scheme has a single critical question that can be used to evaluate an argument fitting the scheme.

The Carneades Argumentation System also supports and uses the argumentation scheme for argument from witness testimony. As shown in Chapter 1 (Section 7), Carneades represents critical questions as different kinds of premises in an argumentation scheme. This approach is made possible, according to the account given in Chapter 1, by drawing a distinction between two kinds of critical questions matching a scheme, based on the notion of burden of proof. One type of premise, called an assumption, automatically defeats the original argument once it has been asked. The other type of premise, called an exception, only defeats the original argument if it is backed up by sufficient evidence that supports it. One could also make this point by saying that in the case of an exception the burden of proof attached lies on the side of the questioner.

The scheme given here for argument from witness testimony is the one contained in the Catalogue of Argumentation Schemes in the latest version of Carneades. *A* is a proposition.

id: witness-testimony
strict: false
direction: pro
conclusion: *A*
premises:

- *W* is in a position to know about things in a certain subject domain *A*.
- Witness *W* believes *A* to be true.
- *W* asserts that *A* is true.

assumptions:

- *A* is internally consistent.

exceptions:

- *A* is inconsistent with the known facts.
- *A* is inconsistent with what other witnesses assert.
- Witness *W* is biased.
- *A* is implausible.

Because the topic of this chapter is on witness testimony that is held to be inadmissible because of suggestions implanted in the mind of the witness through how the witness was later questioned, our concern will be with the third exception, that of bias.

Let's review the general method of the Carneades Argumentation System, to get a preview of how it will be applied systematically to the case in point. As shown in the previous chapters, in a dialogue when participants are engaged in argumentation, proof burdens and standards can be applied. A proof standard is applied as a method for aggregating arguments in an argumentation tree by propagating acceptance or rejection along the branches of the tree. Each argument represents a step from a set of premises to a conclusion, and in many instances an argumentation scheme can be picked from a list and applied to an argument represented as a node in the tree. The argumentation scheme is inserted into the argument node that represents an inference step. The proof standards range in order of strictness from preponderance of the evidence to the highest standard of beyond a reasonable doubt. During a persuasion dialogue of a kind found in a trial setting, the burden of persuasion is set at the opening stage, and is used to determine the outcome at the closing stage. In between there is an argumentation stage during which a burden of producing evidence (burden of production) can shift from one side to the other.

Carneades distinguishes between pro and con argument nodes in an argument tree where the ultimate *probandum* is the root proposition. Weights in Carneades representing the aggregated opinions of an audience can be inserted into an argument tree as real numbers to determine whether a proposition is accepted (in) or not accepted (out). There are three kinds of counterarguments in the Carneades Argumentation System. In a premise attack argument, a con argument is directed against a premise of a prior argument. There can also be an attack against the conclusion of the prior argument. There can also be a Pollock-style undercutter, a con argument attacks the argument itself rather than the premise or the conclusion. Exceptions are modeled in Carneades as undercutters. The latest version of Carneades supports entanglement (Verheij, 2005), an argument structure where an arrow can go from one argument node to another, representing the notion of one argument attacking another and undercutting it. In general, a rebuttal is a counterargument that is directed against a prior argument where the weight of the attacking argument is sufficiently stronger than the weight of the prior argument, in line with the proof standard that is appropriate.

Figure 4.1 shows a simple example of an argument map representing this sort of evidential situation in a case of witness testimony evidence. The +WT notation in the argument shown at the left represents a pro use of the argumentation scheme for argument from witness testimony. This argument is shown as based on the two premises just to the right of it. Underneath, there is a con argument supporting the allegation that Bob's testimony is biased, that acts as an undercutter attacking the argument from witness testimony just above it.

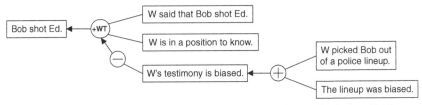

FIGURE 4.1. A Typical Carneades Argument Map of Witness Testimony Evidence

Now we have some idea of how the argument mapping tool of the Carneades Argumentation System is being used along with argumentation schemes in a way that is potentially applicable to issues concerning the fallibility of witness testimony. So we can look at a simpler case that provided a precedent to a more complex case in which it was disputed whether witness testimony of a kind based on how suggestive evidence of a certain sort should be excluded.

2. The Case of the *State v. Classen*

In *State v. Classen* (285 Or 221, 590 P2d 1198 (1979)), the victim hired two men to do yard work, one a black man and the other white. The black man asked to use the toilet, and when the victim reentered her house she found it had been burglarized and both men were gone. She provided a description of the white man of slender build with light brown hair and a Vandyke beard. Seven months later, the victim was shown a photo throwdown with seven pictures. All the men in the pictures had mustaches, but only the defendant had a beard. After being told that the officer suspected that the perpetrator of the crime was among the seven individuals pictured, the victim, with some degree of hesitation, eventually selected the picture of the defendant.

The Classen court ruled that witness identification evidence should not be admissible on the basis of pro and contra arguments. The evidence described as follows is from the summary (10) of Classen presented in the appeal of *Oregon v. Lawson* in the Appellate Court Opinions of the State of Oregon.[2] On the pro side, the court cited the victim's opportunity to view the defendant, "in daylight, under conditions when she could give her attention to him, and without distracting elements of fear or stress," as counting in favor of admission of the evidence. Against this, however, the court noted that the stress of confrontation with a criminal has been cited as making for a high degree of attention that would impress the defendant's picture on the victim's memory.

On the contra side, the court observed that the photo throwdown did not occur for seven months after she had seen the defendant, and that her

---

[2] http://www.publications.ojd.state.or.us/docs/A132640.htm.

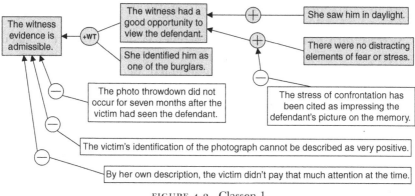

FIGURE 4.2. Classen 1

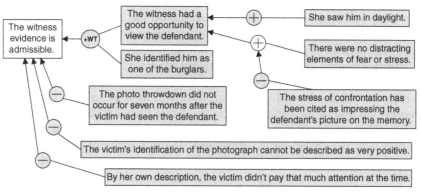

FIGURE 4.3. Classen 2

identification of the defendant could not be described as very positive. The victim said, "at the time I just didn't pay that much attention." Based on these weaknesses, the court arrived at the conclusion that "the state had not carried its burden," and so it ruled the witness testimony inadmissible. Note that in this case, the burden of proof was on the victim to show that the witness testimony was admissible.

In the argument map shown in Figure 4.2, the propositions that are *in* (accepted) are shown in the darkened boxes. In the chain of argumentation at the top, a pro argument from witness testimony is based on the premises that the witness had a good opportunity to view the defendant and the premise that she identified him as one of the burglars. The first premise is supported by two pro arguments. Because none of the four contra arguments shown at the bottom of Figure 4.2 have been accepted yet, they fail to defeat the witness testimony argument shown at the top. Hence, the ultimate conclusion, the proposition that the witness evidence is admissible, is shown in a darkened text box, showing that it is acceptable, based on the evidence displayed in Figure 4.2.

Next we turn to Figure 4.3, where a fuller evidential picture is taken into account. In Figure 4.3, the three contra arguments at the bottom of the argument map are now shown as accepted. Because the mass of evidence represented by these three con arguments at the bottom is strong enough to defeat the network of pro argumentation at the top, the conclusion that the witness evidence is admissible is shown as not accepted in Figure 4.3. To express the evidential situation in Pollock-style terminology, these three arguments at the bottom are not undercutters of the kind shown in Figure 4.1. They are rebutting defeaters, meaning that they are contra arguments against the conclusion of the prior argument. Hence, in Figure 4.3 the conclusion that the evidence is admissible is shown in a white box, meaning it is not accepted.

There is also a complication of Figure 4.3 that needs to be explained. The proposition that the stress of confrontation has been cited as impressing the defendant's picture on the memory is represented as an undercutter of the argument just above it. The question is whether this undercutter is strong enough to defeat the argument above it. This is the situation represented in Figure 4.3, because the node of the argument just above the undercutter has a white background. But as it turns out, it doesn't matter one way or the other whether this undercutter is strong enough to defeat the argument it was directed against. Even if it is not strong enough to defeat it, the three contra arguments at the bottom of Figure 4.3 are strong enough to defeat the argument from witness testimony above them. Without the three contra arguments, the argument from witness testimony is strong enough to prove the conclusion that the witness evidence should be admissible. But once these three contra arguments are brought into the evidential picture, as shown in Figure 4.3, they are strong enough to rebut the prior argument from witness testimony.

Note that so far, the bias issue (for example, the fact that the defendant was the only person with a beard in the pictures) has not yet been taken into account. Now we turn to address it.

3. Scientific Evidence on the Fallibility of Witness Testimony

Witness testimony evidence represents a kind of argumentation that is especially important for modeling legal argumentation, not only because this type of evidence is so extremely common in law, but also because social science research has shown that it is a kind of evidence that can be highly misleading in some instances, and that is very tricky to manage, due to biases and difficulties in processing it in a way to try to minimize such biases.

Loftus (2005, 364) cited a series of experiments showing that misinformation can be planted in the human mind by suggestion. Several studies made up realistic looking but fake Disney ads featuring Bugs Bunny. One poster for the Disneyland resort in California displayed a picture of Bugs Bunny as part of the ad. In several studies, use of a fake Bugs Bunny ad of this sort led 60% of the subjects tested to claim later that they had actually met Bugs Bunny at Disneyland. This is not possible because Bugs Bunny is a Warner Brothers character, and therefore could not be used for copyright reasons in a Disney ad. Findings of this kind raise concerns about how false beliefs can be resurrected from memory by the use of suggestive misinformation. In some follow-up experiments using a fake ad of the same sort, in one experiment 25%–35% of subjects claimed to have met Bugs Bunny, while in another, 62% of subjects said that they remembered meeting and shaking hands with him, and 46% remembered hugging him (Loftus, 2003, 232).

In a useful appendix, the Supreme Court of the State of Oregon in its judgment of the Classen case cited a number of results of scientific research on the reliability of eyewitness testimony relevant to its judgment in this case. The court divided these factors into two categories. Estimator variables refer to characteristics of the witness, the perpetrator and environmental conditions of the event that cannot be manipulated by the agents undertaking the investigation. System variables are factors within the identification procedure that are within the control of the agent's administrative procedure. Under estimator variables the court listed and documented several scientific findings that are important to take into account in evaluating eyewitness testimony that may be subjective.

Under the category of estimator variables were listed viewing conditions. For example, looking at an event under poor lighting conditions can affect the ability of a witness to perceive and remember facts. It was noted that witness confidence can be highly misleading as a bias after the witness has received confirming feedback that he made a correct identification. Under the category of system variables, it was noted, for example, that it can be a source of bias if the witness is not told prior to the identification procedure that it is permissible not to identify anyone. A highly significant finding is that the wording of the question asked by the interviewer can be a highly significant bias factor. Experiments have shown that the memory of a witness can be contaminated by assumptions embedded in the question.

The following seven findings were listed under the category of estimator variables.

1. High levels of stress can significantly impair a witness's ability to recognize faces and keep details and memory.
2. During lineup identifications, it can make a big difference whether the interrogation took place under high-stress or low-stress conditions.

3. Reliability of witness testimony depends on how closely the witness was paying attention.

4. Scientific studies indicated that longer durations of time looking at the perpetrator generally result in more accurate identifications.

5. Viewing conditions, for example looking at an event under poor lighting conditions, can affect the ability of a witness to perceive and remember facts.

6. Physical or mental factors, for example being intoxicated or being very young or very old, are variables that can affect the ability to perceive and remember.

7. Many people tend to assume that witness confidence of certainty is a good indicator of identification accuracy. Some studies have even shown that high witness confidence is the most influential factor in jury determinations of the accuracy of eyewitness testimony. However, other studies show that witness confidence of certainty is highly variable as a parameter for judging accuracy of witness testimony. Confidence can be especially misleading as a bias after the witness has received confirming feedback that he made a correct identification.

The following eight findings were listed under the category of system variables.

1. Administrators of police lineups may imply all kinds of suggestive verbal communication to convey who the suspect is in the lineup.

2. Experts suggest using a person who does not know the identity of the suspect as an administrator of the identification procedure.

3. It can be a source of bias if the witness is not told prior to the identification procedure that it is permissible not to identify anyone.

4. It can be a factor of bias if the lineup is constructed in such a way that the suspect stands out from the other subjects in a way that might lead the witness to select that person.

5. If the witnesses are permitted to view all the subjects together there is a tendency to choose the person who most closely resembles the perpetrator. Such a relative judgment has been found to increase the likelihood of misidentification. To correct this problem, the witness should be presented with the persons or photographs sequentially. Some research, however, challenges the validity of this finding.

6. Presenting the witness with a single suspect shortly after the crime is called a showup. Although showups are widely

regarded as suggestive, and less reliable than lineups, some research indicates that they can be as reliable as lineups.

7. A highly significant finding is that the wording of the question asked by the interviewer can be a highly significant bias factor. Experiments have shown that the memory of a witness can be contaminated by assumptions embedded in the question.

8. Post-identification feedback can inflate the confidence of the witness in the accuracy of his identification. The detrimental effects of this kind of feedback have been well established in the scientific literature on witness testimony bias.

These scientific results about the fallibility of witness testimony have been widely known, both inside and outside law, for over thirty years, and they have been having an impact on police lineup procedures and other methods of processing witness testimony in legal frameworks. One recent highly significant outcome has been a shift in the Oregon courts on how burden of proof is set for eyewitness testimony evidence. Previously in Oregon, the Classen case was the leading precedent. But recently the Oregon Supreme Court, in the case of *State v. Lawson*, 352 Or 724, 291 P3d 673 (2012) shifted the burden to the state to establish that the eyewitness account is admissible.

4. The *State of Oregon v. Lawson*

The ruling in the case of *State of Oregon v. Samuel Adam Lawson*, based partly on legal arguments put forward by the Innocence Network citing evidence from social science research on reliability of witness testimony, changes the way burden of proof will be structured in rulings in the State of Oregon on the admissibility of witness testimony evidence. Previous rulings put the burden of proof on the defendant who wishes to claim that an identification procedure was suggestive. According to the ruling in Lawson, the burden of proof is shifted to the state to prove that the evidence is admissible.

In this case the victims, Noris and Sherl Hilde, were shot with a large caliber hunting rifle in their trailer in a national park in Oregon. She was shot first through the window of the trailer, and then he was shot while trying to use her cell phone to call 911. He died, and she was found lying in the trailer, critically wounded, when emergency personnel arrived. She was taken by helicopter to a hospital, where she was interviewed by police detectives the second day after the shooting. Although she was rambling and hysterical on the way to the hospital, and could not speak well when interviewed by detectives because she was heavily medicated and had a breathing tube in her throat, she was able to give some account of what happened. She was also interviewed by the police two weeks later.

The victims had pitched a tent a week before the incident to claim the campsite. When they arrived at the park with their trailer that day, they found a yellow truck in their parking space and a man occupying the tent. When they told the man it was their tent, he apologized and left, saying that he thought the tent had been abandoned. According to the account at trial given by Mrs. Hilde, after the shooting, the perpetrator had entered the trailer, put a pillow over her face and demanded the keys to their truck, and then walked away. It was possible that she had had a chance to identify the perpetrator at that point.

Samuel Lawson, the owner of the yellow truck, which had been ticketed in the parking lot at the crime scene, admitted he had taken his father's rifle, a .357 Marlin, and ammunition, and that he was camping in the park. He claimed that the rifle had been stolen from his truck. He claimed that he had moved into the Hildes' tent because he believed it had been abandoned.

Part of the evidence was a partial bloody shoe print found near the trailer and a forensic analysis of the fragmented bullets, showing them to be consistent with ammunition of the kind used in the .357 Marlin rifle Lawson had taken to the park. However, the most significant part of the evidence was the witness testimony of Mrs. Hilde. It is important to describe the sequence of the obtaining of this evidence in some detail.

When first asked by the medical personnel who arrived on the crime scene who had shot them, Mrs. Hilde replied "I don't know." When asked later if she knew who did it and whether they were in a vehicle, she replied again that she didn't know. During her ride in the emergency services helicopter Mrs. Hilde sometimes said that the shooter was the man who had been at their campsite, the shooter was the pilot of the helicopter, or that she did not know who the shooter was. However, while being transported in the ambulance, she was informed that her husband was dead. She then identified the shooter as the man who had been in the campground in their tent. Later, during a helicopter ride she told one of the attendants that the shooter had a yellow truck. When the police officer visited her in the hospital and showed her the picture of Lawson, she was unable to identify him as the perpetrator. However, she said that she had seen the perpetrator, that he had been wearing a baseball cap and that he had been in their tent and had a yellow truck. But she added that her eyes were watery and that she was unable to pick out details of the photo. Two years after the crime in a police interview, she identified a picture of Lawson as the man who had entered the trailer, put the pillow on her face and demanded their truck keys. During the trial she testified that Lawson was the man who had shot both her and her husband. She testified that after putting the cushion over her face, and as he walked away, she was able to peek from under the cushion, recognizing him as the man who had been in their tent earlier in the day. When asked if she had any doubt, she responded: "absolutely not. I'll never forget his face as long as I live" (6).

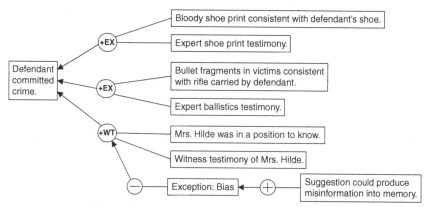

FIGURE 4.4. Carneades Argument Map of the Evidence in the Lawson Case

As shown in the argument map in Figure 4.4, there were two items of expert opinion forensic evidence – a bloody shoe print and the matching of bullet fragments found in the victims consistent with the type of bullets used in the rifle carried by the defendant to the park. These two separate arguments are shown at the top of Figure 4.4. Below them is shown the argument from witness testimony based on the testimony of Mrs. Hilde. Mrs. Hilde identified her attacker several times during the sequence of collection of evidence by the police, but the argument map shown in Figure 4.4 is simply an overview representing the main features of the evidence in general outline. The forensic evidence was weak and questionable in certain respects, and so whether the burden of proof to show that the defendant committed the crime beyond reasonable doubt was met depended on the eyewitness testimony evidence.

5. The Change Made in Oregon Law

In the case of *State v. Classen*, the Oregon court had recognized the unreliability of eyewitness testimony under suggestive circumstances by setting in place a procedure to decide the admissibility of such evidence. A defendant who wants to suppress eyewitness testimony evidence on the grounds that it is the product of a suggestive procedure must carry out two steps. First, it must be determined whether the process leading to the identification is suggestive, and second, independent evidence must be provided that substantially excludes the risk that the identification resulted from the suggestive procedure. The second determination needs to be made, according to this test, by five factors: the opportunity the witness had to get a clear view of the person, the attention the witness gave to the identifying features of the person, the timing and completeness of the description given by the witness, the certainty expressed by the witness in making the identification and the lapse of time between the original observation and the subsequent identification.

The Oregon Supreme Court ruling made in the Lawson case shifted the burden of proof from the defendant to the prosecution, the state in a criminal case: "the state as the proponent of the eyewitness identification must establish all preliminary facts necessary to establish admissibility of the eyewitness evidence" (44). At a minimum, the state must prove that the eyewitness has personal knowledge of the matters to which the witness will testify, and must have proof that the identification is rationally based on the witness's first-hand perceptions and helpful to the trier of fact. If the state satisfies its burden that the witness testimony evidence is admissible by this criterion, the burden shifts to the defendant to prove that the probative value of the evidence is outweighed substantially by the danger of unfair prejudice, or by other factors including misleading the jury or undue delay (44–45).

Although there has been considerable recent research on modeling burden of proof in the field of AI and law, there has not been an example analyzed yet and modeled with the current tools that concerns burden of proof for admissibility of evidence. Three types of burden of proof applicable during a legal trial (along with two other less generally significant types not considered in this chapter) have been identified by Prakken and Sartor (2009): burden of persuasion, burden of production and tactical burden of proof. They note (2009, 227) that the distinction between burden of production and tactical burden of proof is usually not clearly made in common law, or even explicitly considered in civil law countries, but is important for both systems because it is induced by the logic of the reasoning process. So far, burden of proof for admissibility of evidence has not been considered or studied in the AI and law literature on burden of proof. What is indicated is that a new kind of burden of proof has to be recognized – burden of proof for admissibility of evidence.

A number of factors were listed as applying to the witness testimony evidence during the Lawson case. Not only was Mrs. Hilde subject to high levels of stress and fear when she observed the perpetrator, but it was dark inside the trailer and her view was only partial because her face was covered by a pillow. It was noted that memory decays over time, and that her identification of the defendant took place two years after the crime. There are many procedural questions about the police interviews of Mrs. Hilde. She was heavily medicated when first interviewed in the hospital. The police questioned her using leading questions. She was susceptible to memory contamination from repeated suggestive questioning. The police planted the suggestion in her mind that the man she saw earlier at their campsite was the perpetrator. During two photographic lineups she was unable to identify the defendant. It was only after she had seen a newspaper article with a picture of the defendant was she able to identify him.

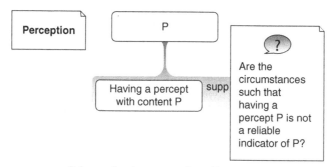

FIGURE 4.5. Scheme for Argument from Perception in Rationale

The sequence can be described as a process of suggestion and confirming feedback.

6. Reconfiguring the Argumentation Schemes

In this section, the argumentation scheme for argument from witness testimony is reformulated. The way of doing this pays careful attention to the relationship between the scheme for argument from witness testimony and the scheme for argument from perception. We start with Pollock's version of the scheme.

Pollock's version of the scheme for argument from perception can be formulated, modifying the variables slightly to match the schematic formulation, as follows (Walton and Sartor, 2013).

> Premise 1: Agent a has a P image (an image of a perceptible property).
> Premise 2: To have a P image (an image of a perceptible property) is a
> *prima facie* reason to believe that the circumstances exemplify P.
> Conclusion: P is the case.

There is only one critical question attached to this scheme, CQ1: are the circumstances such that having a P image is not a reliable indicator of P?

As shown in Figure 4.5, the scheme for argument from perception is represented in Rationale in an even simpler way with only one premise and one critical question. Rationale is domain-independent, but has been applied to legal argumentation (van Gelder, 2007).

The comparable scheme in Carneades is called argument from appearance, based on the account of that scheme given in the argumentation literature (Walton, 2006a).

> id: appearance
> strict: false
> direction: pro

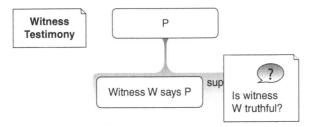

FIGURE 4.6. Scheme for Argument from Witness Testimony in Rationale

conclusion: *O* is an instance of class *C*.
premise: *O* looks like a *C*.

No critical questions are given in Carneades matching this scheme, but perhaps one similar to the critical question in Rationale could be given, asking whether there are circumstances suggesting that the appearance is not a reliable indicator that *O* is an instance of *C*.

The scheme for argument from witness testimony in Rationale is shown in Figure 4.6. Here again the scheme is expressed in a simple way. There is only one critical question for it.

The scheme for argument from witness testimony in Carneades shown in Section 1 was based on a comparable scheme formulated in Walton (2008, 52) as follows.

> Position to Know Premise: Witness *W* is in a position to know whether *A* is true or not.
> Truth Telling Premise: Witness *W* is telling the truth (as *W* knows it).
> Statement Premise: Witness *W* states that *A* is true (false).
> Conclusion: Therefore (defeasibly) *A* is true (false).

We have seen that the examples of argument from witness testimony modeled in Carneades described earlier do not make any use of the truth-telling premise. They adopt a simpler version of the scheme with only two premises, which could be expressed in the following form.

> Position to Know Premise: Witness *W* is in a position to know whether *A* is true or not.
> Statement Premise: Witness *W* states that *A* is true (false).
> Conclusion: Therefore (defeasibly) *A* is true (false).

This way of formulating the scheme is recommended here as the one that should supplant the other versions in Carneades, Rationale and other argumentation systems. It is more complex than the Rationale version because (like the Pollock scheme) it has two premises instead of one, but is less complex than the Carneades version, which had the premise that the witness believes the proposition *A* to be true. The Pollock version makes the scheme simpler and easier to apply, while the role played by the truth-telling premise can now be taken up by the additional critical questions considered in the next section.

In this version, the scheme for argument from witness testimony can be seen as based on the scheme for argument from perception because the latter type of argument, citing the perception of the data by the witness, is the basis for claiming that the witness is in a position to know about the claim at issue.

7. The Critical Questions Matching the Scheme

In the Walton account, issues relating to bias are subsumed under the trustworthiness critical question, a more general critical question, and there is no specific bias of question. It is suggested here that the Carneades formulation of the scheme is better in this respect, given that this chapter has emphasized the importance of the role of bias in evaluating witness testimony evidence, and that there is specifically a bias subquestion in the Carneades version of the scheme. But this scheme could be simplified by dropping one premise.

Now we come to the formulation of the critical questions matching the scheme for witness testimony. In the lawsuit case, there were a lot of internal inconsistencies in the testimonies given by Mrs. Hilde at various times between the crime and the trial itself. This was not in itself the main problem, however. The main problem was the suggestive nature of the police procedures of questioning, and this consideration fits best under the bias question. The question here is whether the witness was biased as she was questioned by the police, shown pictures of the supposed perpetrator and so forth. But notice that the bias question is classified in Carneades as an exception. This means that whoever claimed that the witness is biased, in order to exclude the witness testimony, has the burden of proof to back up this allegation with evidence. This ruling on burden of proof was applicable in the Classen case. However, in the Lawson case the burden of proof was reversed, and the burden now fell on the prosecution to establish the reliability of the witness testimony.

Once we realize that argument from witness testimony is built on the two foundational schemes of argument from perception and argument from memory, we can see that there is a timeline joining the three schemes in an evidential sequence.

How the three schemes are nested together is shown by the timeline in Figure 4.7 at the top. First, there is the perception of the event, including the identification of the suspect in a criminal case, by the witness. Second, there is the trace of this event in the memory of the witness. Third, this memory is carried forward when the witness offers testimony concerning the event or the identity of the suspect. Each argumentation event is founded on the

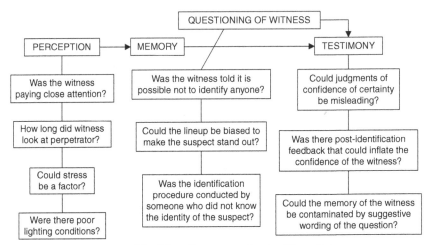

FIGURE 4.7. The Time Sequences for the Three Schemes

other, starting with perception, going to memory and from there to the witness testimony. It also needs to be recognized that there is an interval that is very important from an evidential point of view – between the memory stage and testimony stage. This stage is the sequence of events during the questioning of the witness. For example, in a criminal investigation it would consist of a police officer showing the witness pictures of the suspect and asking the witness questions about whether the person in the picture is the same individual as the suspect. Or it could consist of a police lineup where the witnesses are asked to identify one of the persons in the lineup as the suspect.

Underlying each of the three schemes is a set of critical questions that applies to the scheme or procedure above it. These critical questions have been abstracted from the list of seven estimator variables and the list of system variables described in Section 3. The critical questions in the column on the left are attached to the argumentation scheme for argument from perception. The critical schemes in the middle column relate to the procedure for the questioning of the witness in order to access the memory of that witness to obtain testimony that can be used as evidence. The critical schemes in the right column are attached to the argumentation scheme for argument from witness testimony. All these critical questions are best regarded as subquestions of the main questions in the argumentation scheme. The critical questions in the middle column and the right column come under the critical question of bias as a subquestion of the scheme for argument from witness testimony in the Carneades system.

In the Carneades Argumentation System, it still appears to be best to continue to classify the bias question as an exception, because normally, if one party alleges the second party is biased in order to refute the second party's argument, the argument has little force unless it can be backed up with some specific evidence of bias. Otherwise the first party can simply reply, "Of course I'm not biased. If you think I am, then it is up to you to get some of this evidence to support your allegation." However, once the requirements for admissibility set by the Lawson court are applied, the burden shifts to the other side to show the relevance of the allegation of bias. But how to deal with the relevance of critical questions matching the argumentation scheme is not an issue that has even been raised in the literature on argumentation, as far as I am aware. So the implications of this new way of dealing with bias being recommended by the Oregon Supreme Court have not yet been explored in connection with the scheme for argument from witness testimony. These implications are drawn out in the next three sections.

8. Admissibility, Bias and Burden of Proof

One conclusion suggested by the analysis of witness testimony that emerged from the discussion of the ruling in the Lawson case is that the previous types of burden of proof recognized in the AI and law literature need to be supplemented by a new type of burden of proof for admissibility. Carneades had previously recognized the distinction between burden of persuasion and burden of production, as well as the other types of burden of proof recognized by Prakken and Sartor. But it would appear from the examples studied in this chapter that burden of proof for admissibility is a different type of legal burden. How it could be fitted into the Carneades framework with its three stages of dialogue is indicated in Figure 4.8.

As shown in Figure 4.8, burden of proof for admissibility of witness testimony evidence would normally be considered in pretrial discussions about admissibility, but the issue of admissibility could also be raised during argumentation stage as the trial is underway. From a point of view of argumentation theory, what has been shown is that there is one kind of burden of

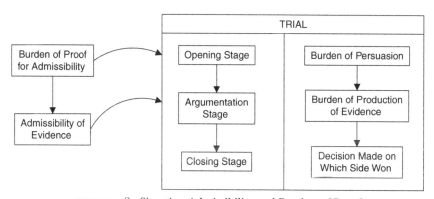

FIGURE 4.8. Situating Admissibility and Burden of Proof

proof that an argument has to meet in order to be a successful proof of its conclusion, but there is also another kind of burden of proof that has to do with reliability of evidence.

An area for further research is the connection between the two categories of variables now established in scientific research on evaluating witness testimony, estimator variables and system variables. It could be conjectured as a hypothesis that the estimator variables could be studied in relation to the schemes for perception and memory. Following this approach, the system variables, representing the different kinds of bias that might be encountered in police procedures for collecting eyewitness evidence, in contrast, can be studied in relation to the scheme and set of critical questions for argument from witness testimony.

Witness testimony is also important as evidence in other fields, for example in history, but because there are special rules in law that have evolved to deal with it, it would seem reasonable that modeling the argumentation scheme for argument from witness testimony along with its set of critical questions should have a special format for legal applications.

Another conclusion arising from this chapter is that the argumentation scheme for argument from witness testimony needs to be modified. In this chapter, the Carneades Argumentation System was used (along with scientific findings about the fallibility of eyewitness testimony) to provide analyses of the evidential situations in two cases in which the law was changed in Oregon. In the second case, the Oregon Supreme Court shifted the burden of proof to the state (the prosecution) to establish that eyewitness evidence is admissible. The specific conclusions of this chapter consist in the reformulations of the scheme and critical questions for argument from witness testimony, based on the need to manage suggestion biases studied via the use of the two cases from the Oregon courts.

9. Admissibility, Relevance and Examination

Chapter 8 confronts the question of whether the legal way of managing burden of proof in trials has features that are also applicable to the management of burden of proof in everyday conversational argumentation. In this connection, what the lessons are for the management of argumentation schemes and critical questions that were learned in Chapter 4 have to be determined in light of the differences between the two contexts of dialogue. In law there are procedural rules that are specifically laid out before the trial begins, and that govern many of the matters connected with argumentation as the trial proceeds. Relevance will turn out to be a vitally important concept in Chapter 8. One important feature of the legal framework of the trial is how the notion of relevance is defined and regulated.

Relevance is defined in the American Federal Rules of Evidence (FRE) in Rule 401. Relevant evidence is considered to be any "evidence having any tendency to make the existence of any fact that is of consequence to the determination of the action more probable or less probable than it would be without the evidence."[3] What does this mean in plain English? The "action" refers to the ultimate *probandum* to be proved (or cast into doubt) in the given case. It could also be called the ultimate claim that has to be proved. An argument considered as evidence in a trial is relevant if it gives more probative weight or less probative weight to the ultimate claim in the case. How probative weight connects to relevance can best be represented using a tree structure comparable to a typical argument graphs used in a Carneades argument diagram. There needs to be a central claim, often called an ultimate *probandum* in law, at the root of the tree. An argument node *A* in such a graph is relevant to another argument node *U* (representing the ultimate claim) if and only if there is a path from *A* to *U* in the graph. On this model, an argument is relevant, in both the legal and everyday sense, if it can be chained forward along a path that leads to the ultimate claim in a given case. If *A* has probative weight, it will boost up or lessen the probative weight of the ultimate claim. To use the Carneades Argumentation System to make a determination of relevance, the arguer needs to begin by providing input on which premises the audience has accepted or not. Then the system could potentially be used to search for a path leading from these premises (along with others) to the ultimate claim.

In law, evidence is generally admissible in a trial only if it is relevant. The general rule is that evidence is taken to be relevant, meaning, if it is taken to have the capability of altering the probative weight of the ultimate claim in the case being tried. There are important exceptions to this general rule, however. For examples, another FRE rule (403) excludes evidence that might have some probative value, but might also tend to prejudice the jury. In these kinds of cases, the weak probative value of the evidence is outweighed by the prejudicial effect it might have if admitted into the trial. In general, according to this rule, evidence may be excluded "if its probative value is substantially outweighed by the danger of unfair prejudice, confusion of the issues, or misleading the jury, or by considerations of undue delay, waste of time, or needless presentation of cumulative evidence."

Another way that the trial setting of argumentation is quite different from the setting of argumentation in everyday conversational settings concerns the way witness testimony is questioned and evaluated. In everyday conversation argumentation, it may be possible to question a witness who has made a claim in some cases, but in many cases this may not be possible at all. In a trial, it is important for witness testimony to be subject to critical questioning of a kind called examination (Walton, 2008, 210–243). To avoid the criticism that what was claimed to have been said is merely

---

[3] http://www.law.cornell.edu/rules/fre.

hearsay, it is important that the witness should appear in the trial, and be available for questioning. When the lawyer for one side critically questions the testimony of a witness brought to testify by the opposing side, the dialogue is called cross-examination. From a point of view of argumentation theory generally, examination dialogue can be classified into two subtypes (Walton, 2008, 213–214).

Examination dialogue is the testing vehicle used for evaluating cases of argument from expert opinion. The examinee needs to ask the appropriate critical questions matching the scheme for argument from expert opinion, but as the process continues he may need to probe even more deeply into the testimony of the expert in order to follow up his initial questions and to ask for explanations of what was said. The purpose of the dialogue can be called one of exegesis, a process in which the examiner constructs explanations of the text to find the supposed meaning of what the expert has said. The process is one of inference to the best explanation. The analyst produces a hypothesis about what he takes the text to mean. Then the evidence of the text can be used to argue that the hypothesis is stronger or weaker, and additional hypotheses can be formulated. There are three levels in an examination dialogue (Walton, 2006b, 759).

1. Collection of Data. Examination begins by looking at something and collecting information.
2. Testing. The information collected is tested to see if it can be confirmed by a test.
3. Hypothesis Formation. The findings of the first two levels are critically discussed, evidence for or against competing hypotheses are evaluated and a best explanation is selected.

The argumentation at the third level is also part of the testing, like that at the second level, but it is highly argumentative.

It is easy to see that the way the law deals with witness testimony, argument from expert opinion and other related forms of argumentation are quite different in key respects from the way such arguments are evaluated in everyday conversational argumentation. Even so, important lessons can be learned from the way witness testimony evidence was shown to be dealt with in legal argumentation in a trial setting in Chapter 4. In Chapter 8 it will be shown how these lessons are transferable to the study of everyday argumentation outside the legal setting. Both settings share a dialogue structure that has the characteristic three stages, an opening stage, an argumentation stage and a closing stage.

10. Conclusions

The general conclusion of this chapter concerns the theoretical question in argumentation theory of whether there should be a burden of proof for relevance. This question is one with broad implication for argumentation studies, both within and outside law. There are many traditional informal fallacies, such as appeal to pity, appeal to fear, *ad hominem* argument and so forth, that are in many instances inherently reasonable arguments that carry probative weight, but that become fallacious when they are irrelevant. Judicial wisdom, reflected in the Federal Rules of Evidence in particular, can offer a basic notion of relevance that can be extremely useful for providing some clear, precise and systematic method for judging relevance of an argument, or other type of speech act like the asking of the question, in any given case. These observations take us to a next level, the level where a problematic case concerning burden of proof needs to be decided by a court by changing the currently accepted rule on how burden of proof should be managed. The management of burden of proof decisions at this higher level requires a capacity for reasoned argumentation at a higher level about applying conflicting rules for burden of proof allocation to a case where assigning burden of proof to the case is itself a disputed issue.

Evidence is probatively relevant if it can be used to prove or disprove the ultimate contention (thesis, opinion) of one side or the other. But now what does this definition of relevance tell us about the information that is supposed to be elicited in examination of a witness? It tells us that there is an underlying notion of logical relevance shift involved in the trial as a dialectical procedure. The information-seeking dialogue that takes place in examination of a witness needs to be judged relevant or irrelevant on the basis of whether that information can be used to prove or disprove the ultimate contention of one side or the other in the trial as a whole. The trial as a whole is not an information-seeking dialogue. The trial is a persuasion dialogue, a type of critical discussion in which the goal is to resolve an initial conflict of opinions (Feteris, 1999). Everything in the trial is subservient to this ultimate goal. So is the information-seeking dialogue in the examination of a witness. The goal of this dialogue is to elicit information. But it is not to collect just any information. The goal is to elicit information that can be used in resolving the conflict of opinions that is the main issue of the trial. Thus, the information-seeking dialogue is embedded in the persuasion dialogue.

When you look at the argumentation underlying it, legal evidence consists of two factors. First, there must be premises that provide the information relevant in a case. Second, there are the conclusions that need to be drawn from these premises, but assurance is required that the information presented in witness testimony is reliable. Testimony that is biased in the way illustrated by the Lawson case is not reliable.

In the Lawson case the Oregon Supreme Court shifted the burden of proof to the state to prove that witness testimony must meet certain requirements in order to be admissible as evidence. The burden of proof was shifted to the state to prove that the eyewitness must have personal knowledge of the matters to which the witness testified, and must have proof that the testimony is based on the witness's first-hand perceptions. Only if this burden is satisfied, according to the court's ruling, is witness testimony evidence admissible in a given case. Once the state has satisfied this burden, the burden then shifts to the defendant to prove that the probative value of the evidence is outweighed by the danger of unfair prejudice. This change in Oregon law acts to affect future cases where eyewitness testimony is a part of the evidence in the Oregon courts. What is shown in cases of this kind is that in law, from time to time, rulings are made in a general way about how burden of proof is to be treated.

In Chapter 5 it will be shown how the notion of relevance of argumentation needs to be decided by fitting the argument into a context of dialogue, a procedural framework for uses of argumentation. In light of this theory, which shows how burden of proof works in formal dialogue systems, evidence such as that supplied by witness testimony in a case breaks down dialectically into two components. First, there is the information presented as an assertion or claim (a proposition). The information is presented in the form of the witness testifying that she saw or heard something, or was otherwise in a position to know about something she can now relate to the court. Second, there is the procedure of drawing conclusions from this information. To qualify as relevant evidence an argument must have probative value in relation to what needs to be proved in the case. In every legal case, there is an issue, or "action," a conflict of opinions that is supposed to be resolved by the trial. Whatever information or logical inference that can be used to support or raise critical questions about the contention of one side or the other of the conflict of opinions is relevant. "Relevance" is meant here in the dialectical sense of referring to anything that carries probative weight that can be used to help resolve an initial conflict of opinions. The legal issue in the Lawson case was not specifically one of relevance, but was one of admissibility of witness testimony. Even so, in argumentation theory generally, the problem of whether an argument is admissible, or should be considered too weak or prejudiced to be considered as worth taking into account, is one of relevance. What has been revealed then is that questions of relevance are linked to questions of burden of proof in some significant cases.

5

Burden of Proof in Dialogue Systems

In his book on fallacies, Hamblin (1970) built a simple system for argumentation in dialogue he called the Why-Because System with Questions. In his discussion of this system, he replaced the concept of burden of proof with a simpler concept of initiative, which could be described as something like getting the upper hand as the argumentation moves back and forth in the dialogue between the one party and the other. No doubt he realized that the concept of burden of proof was too complex a matter to be dealt with in the limited scope of his chapter on formal dialogue systems. In this chapter, it is shown how an extended version of Hamblin's dialogue system provides a nice way of modeling the phenomenon of shifting of burden of proof in a dialogue, yielding with a precise way of distinguishing between different kinds of burden of proof, and dealing with fallacies like the *argumentum ad ignorantiam* (argument from negative evidence).

Over forty years has passed since the publication of Hamblin's book *Fallacies* (1970), and there has been much written on the subject of argumentation since that time. One might think that such a book would have long ago ceased to have much value in contributing to the latest research. Such is not the case, however, especially with regard to Hamblin's remarkably innovative Chapter 8 on formal dialogue systems, a chapter that provided the basis for much subsequent work. To give an example of a formal dialogue system of the kind he recommended in Chapter 8, he built a Why-Because System with Questions. A leading feature of this system is that it has a speech act representing a move in a dialogue in which one party asks the other party to prove, or give an argument to support a claim made by the first party. The Hamblin system has several rules for managing dialogues in which such support request questions are asked, and need to be responded to. It is shown in Chapter 5 how these rules are fundamentally important in attempting to build any formal dialogue system designed to be a framework modeling the operation of burden of proof in rational argumentation.

A simple example of a seven-step dialogue is examined in which one party tries to avoid taking his proper turn by making a clever move that backfires. The dialogue contains an attempt to evade burden of proof of the kind that has been associated with the *ad ignorantiam* fallacy. It also turns out to be an interesting example for testing the extended version of Hamblin's system. Chapter 2 explained sophisticated systems for analyzing burden of proof in legal argumentation. However, one problem is that now we have such a multiplicity of different complex systems for modeling different kinds of dialogues in different special contexts (for example legal argumentation and scientific argumentation), that there is a tendency to lose sight of the basics. One can see from the way Hamblin designed the system to be simple, basic and flexible that it could be adapted to different kinds of rules about burden of proof that need to be formulated in different ways to be tailored to the specific needs of a given procedural context. The puzzles are solved in this chapter by extending Hamblin's system to set up the basic kinds of moves in a dialogue in a general way that extends to the modeling of burden of proof in many different argumentation settings.

In Section 1, a summary of the basic structure of the Why-Because System with Questions is given, and in Section 2 a simple example of an everyday conversational dispute involving burden of proof is presented. The argumentation in the example is analyzed in Section 6 and evaluated in Section 7, using the new dialogue system. The solution to the problem is given in Sections 7, 8 and 9. The conclusions are summarized in Section 10.

1. Hamblin's Why-Because System with Questions

Hamblin (1970, 265–276) built a simple dialectical system, the Why-Because System with Questions, designed to show that problems of organizing commitments can be solved. There are two participants called White and Black. By convention, White moves first, and then the two parties take turns making moves. The language is that of propositional calculus, but it could be any other logical system with a finite set of atomic statements (265). As each party moves, statements are either inserted into or retracted from the commitment set of the party who made the move. A record of each party's commitments is kept throughout the dialogue and updated at each move. On Hamblin's definition, "a speaker is committed to a statement when he makes it himself, or agrees to it as made by someone else, or if he makes or agrees to other statements from which it clearly follows" (Hamblin, 1971, 136). In Hamblin's view, a commitment is not necessarily a belief, although a speaker's commitment to a proposition can often be an important indicator that he or she believes that this proposition is true. Acceptance can be treated for our purposes in this chapter as equivalent to commitment. Commitment is a function of the moves each party has made in a dialogue.

At each move in a dialogue, a participant is allowed to say various things called locutions by Hamblin, but nowadays we would call them speech acts used in a dialogue. The names given below for the types of locutions are mine, but reflect Hamblin's intent. In this chapter a careful distinction is drawn between the notions of statement and assertion. The concept of a statement will be taken to be equivalent to the concept of a proposition. Only propositions, or statements if you like, are the bearers of truth values. An assertion is treated as a kind of speech act. More precisely, the making of an assertion is described as an action taking place in a dialogue when a participant puts forward an assertion as a claim made. In this chapter, therefore, the speech act of going forward with an assertion will often be described equivalently as the making of a claim by asserting a particular proposition. An assertion, in other words, has three elements: the party who made the assertion, the proposition that was asserted, and the move in an orderly dialogue at which the assertion was made.

Assertion: 'Assertion A' is the speech act of putting forward a statement. When a party asserts a statement, it goes into his commitment set. In special instances a party can also say 'Assertions A, B.'

Retraction: 'No commitment A' is the speech act of retracting a commitment, assuming that the party was previously committed to A. If he was not committed to A when he said 'No commitment A,' he could simply be making it clear that he is not committed to A, even though in the simplest dialogue system of this sort, both parties can see all the statements in both their own commitment set and the other party's.

Yes-No Question: 'Question A, B, ... Z' is the speech act of asking whether the hearer thinks that selected statements are true or not.

Support Request: 'Why A' is a request for the other party to supply an argument that would give reason for him to accept A. Such an argument needs to have A as its conclusion and it needs to have one or more premises.

Resolution Request: 'Resolve A' is a request for the addressee to make clear where he stands with respect to some instance where he has committed himself to both A and not-A.

This last type of move is important for Hamblin, as he is interested in modeling a Socratic-style discussion where the questioner leads the respondent to commitment to an inconsistency.

Hamblin defines his general notion of a dialogue containing moves and locutions more precisely in his 1971 article. He begins (130) with a set of participants P and a set of locutions L. He defines a locution-act, which amounts to a speech act used by a participant in the dialogue, as a set of participant-locution pairs (1971, 130). For example, $\langle P_0, L_4 \rangle$ is a locution act where P_0 is the first participant and L_4 is the fourth type of locution allowed in the dialogue. For example, L_4 may be the asking of a why-question. A dialogue of length n is defined as a member of a set of sequences of locution acts. He illustrates this definition by giving an example of a small dialogue of length 3: $\{\langle 0, P_0, L_4\rangle, \langle 1, P_1, L_3\rangle, \langle 2, P_0, L_2\rangle\}$. In this example, participant P_0 starts the dialogue at move 0 by uttering a locution of type 4. At move 1, participant P_1 replies by putting forward a locution of type 3. At a move 2, participant P_0 replies using a locution of type 2. Generally a dialogue is an ordered sequence of moves of this sort. In Hamblin's view, how any particular type of dialogue is defined depends on what locutions are allowed and how these locutions or speech acts are defined.

For our purposes, as noted earlier, we can treat the speech act of making an assertion as equivalent to the act of making a claim. The important things about making a move fitting this speech act are that (1) it commits the speaker to the statement made, and (2) it represents a strong form of commitment that commits the speaker to defending the claim, if asked to do so (Walton and Krabbe, 1995). So for our purposes we can work with what we will call a Why-Because System (WB System), a simpler system that has only assertions, retractions and support requests, but that can be made more complex by the addition of other speech acts and rules.

Hamblin (1970, 166) also has a number of syntactical rules for his Why-Because System with Questions. One of these rules is especially significant. When simplified into a form suitable for a Why-Because System, it is the rule that when one party asks the question 'Why A,?' the other party must reply by putting forward one of the following three speech acts: Assertion A; No commitment A; Statements B, $B \rightarrow A$ (where $\rightarrow$ represents the material conditional of propositional calculus). Let's call this rule the three responses rule. It is this particular rule that appears to be related most closely to the notion of burden of proof. However, it is not the same thing as the standard rule for burden of proof that requires any party who has made a claim to back up that claim with support if challenged to do so by the other party in the dialogue. It is a different rule because it allows the party to whom the why question is addressed the two other options of saying 'Assertion A' or 'No commitment A.'

This rule also brings in a number of other complications in that it relates to two other rules for formal dialogue systems that Hamblin (1970, 271) considers, even though he does not require them as mandatory rules for the Why-Because System with Questions. One is the rule that 'Why A?' may not be asked unless A is a commitment of the hearer and not of the speaker. This rule would obviously affect the three response rule. Indeed it would even conflict with it, because there is no need to allow the replies 'Assertion A' and 'No commitment A' if 'Why A?' may only be asked if A is not a commitment of the hearer. The second rule relates to the support answer to a why question, and it relates to the commitments of the two parties. This rule (Hamblin, 1970, 271) requires that the answer to a why question, if it is not 'Assertion A' or 'No commitment A,' must be in terms of statements that are already commitments of both speaker and hearer. Let's call this rule the Commitment to Premises (CtP) Rule. Hamblin does not advocate CtP. Indeed, he describes it as "an unnecessarily strong rule" (271). However, it is useful to take this rule into account, because it closely relates to the support request speech act for the Why-Because System formulated earlier, as will be shown when we go on to discuss how to more precisely formulate this rule.

2. An Example Argument

The following dialogue is a disputation between two parties, Alfred and Dana, on the issue of whether Bob stole Kathy's garden rake. Alfred has made the allegation that Bob stole Kathy's rake. Dana claims that Bob did not steal Kathy's rake. Thus, there are two sides to the dialogue, and each side makes a claim that is the negation of the claim put forward by the other side. The two parties take turns engaging in argumentation. The seven moves in the dialogue are shown in Table 5.1. Dana opens the dialogue at move 1 by asking Alfred to prove that Bob stole Kathy's rake. He poses a why question asking Alfred to prove his claim. At move 2, Alfred responds by offering some evidence to support his claim. Alfred replies that Bob took the rake from Kathy's yard. At move 3, Dana follows up with another why question, asking Alfred to support his assertion made at move 2. At move 4 Alfred responds to Dana's request by offering some evidence that supports his previous claim that Bob took the rake from Kathy's yard. He offers some witness testimony, saying that a third party, Mary, saw Bob take the rake from Kathy's yard.

Up to move 4, Alfred seems to be winning the argument. At move 5, however, Dana puts forward an argument that attacks Alfred's argument made at a move 4. This argument may not be strong, however, because Alfred could easily respond to it in various ways. For example, he could argue that even though Mary has lied in the past, that fact is not a good reason to think that she might be lying in this instance. Or he could question whether Mary has lied in the past, and challenge Alfred to prove that claim. But instead of making either of these moves, Alfred has taken a radically different step in the dialogue by asking Dana to prove that Bob did not steal Kathy's rake.

This move can be described as an attempt to shift the burden of proof to the other side. Some might say that this move is improper, even amounting to committing of the fallacy of argument from ignorance (the *ad ignorantiam* fallacy), because Alfred is merely trying to avoid taking his proper turn by responding to Dana's previous argument that Mary has lied in the past. It seems to be a clever move, but in this instance it backfires. For at move 7, Dana makes the surprising claim that the rake was not Kathy's property. At move 7, when Alfred asks Dana to prove this claim, Dana replies with an

TABLE 5.1. *The Rake Theft Dialogue*

Dana	Alfred
1. WHY [Bob stole Kathy's rake]?	2. Bob took the rake from Kathy's yard.
3. WHY [Bob took the rake from Kathy's yard]?	4. Mary saw Bob take the rake from Kathy's yard.
5. Mary has lied in the past.	6. WHY [Bob did not steal Kathy's rake]?
7. Bob has a bill of sale showing he bought the rake.	

argument that could still be open to critical questioning or attack, but in the absence of a convincing refutation looks like persuasive evidence.

The rake theft example is only a very simple one, made up of seven moves. But it has three Hamblin-style why questions among the seven moves, and the dialogue presents some other interesting features because it contains what appears to be an attempt to evade burden of proof of the kind that has been associated with the *ad ignorantiam* fallacy. It contains other interesting features of argumentation, as will be shown, and it will turn out to be an interesting specimen for us to try to analyze using the tools presented in Hamblin's Why-Because System with Questions. It is important to note that the example is not an instance of legal argumentation, but looks similar in outline to the kind of argumentation that could take place in a criminal case of theft. Is not meant to represent a case that has gone to court, or where a criminal charge has been made.

3. Burden of Proof in Dialogue

In Hamblin's Why-Because System with Questions, any assertion made by either party can be challenged, and when an assertion is challenged, the party who made the assertion is obliged to either prove it at his next move or give it up. This way of handling burden of proof is common in many other approaches. The rule that when challenged to defend an asserted proposition, one must either defend it or else retract it as widely, but not universally, held by philosophers (Rescorla, 2009a, 87–88). Some philosophers, for example Brandom (1994, 177), claim that there are exceptions like the propositions "There have been black dogs" and "I have ten fingers."

The rule governing burden of proof in (van Eemeren and Grootendorst, 1992, 208) requires that "a party that advances the standpoint is obliged to defend it if the other party asks him to do so." This rule initially appears to be similar to rule 8a of the dialogue system PPD of Walton and Krabbe (1995, 136), which says, "If one party challenges some assertion of the other party, the second party is to present, in the next move, at least one argument for that assertion." There may be important differences between these two rules, however, once we try to specify more precisely what each rule is intended to do in a dialogue system.

The concept of formulating a standpoint in a critical discussion refers to the initial conflict of opinions set in place at the opening stage of the dialogue where the fundamental issue of the dialogue is stated and agreed upon by both parties. When we say that the fundamental issue of the dialogue has to be stated and agreed upon by both parties at the opening stage, we are stating that the unsettled issue to be discussed has to be formulated in order for normative judgments to be made on matters like whether an argument is relevant. In other words, the parties must agree on what the dialogue is supposed to be about. It is possible, nevertheless, to have discussions between parties who disagree even on what the fundamental issue of the dialogue should be. This kind of discussion needs to take place at the opening stage, and what the issue is needs to be settled at that stage before the dialogue can properly proceed to the argumentation stage. There is also an interesting kind of exception becoming evident. The dialogue can shift to a different level called a metadialogue (Krabbe, 2003), in which the parties, perhaps assisted by a mediator, a judge or some other third party, can sort out procedural matters, for example, whether the issue was correctly formulated at the opening stage. Another problem that sometimes needs to be sorted out by shifting to a metadialogue is the burden of proof. However, in this chapter, we shall be exclusively concerned with problems of the shifting of burden of proof that take place during the argumentation stage itself, and where no shifting to a metadialogue is being considered.

Given that the fundamental issue of the dialogue has been stated and agreed upon by both parties at the opening stage, the rule that the party is obliged to defend its standpoint if the other party asks him to do so seems to refer to a kind of burden of proof set at the opening stage that then governs the various moves that are made during the argumentation stage. It is helpful now to explore a broader distinction that applies, not only in legal argumentation, but that can be applied to conversational argumentation generally. This is the distinction between global and local burdens of proof (Walton, 1988). Global burden of proof is set at the opening stage of a dialogue, applies through the whole argumentation stage, and is used to determine which side was successful or not when a ruling needs to be made when it is determined who won or lost at the closing stage. In contrast, local burden of proof applies to speech acts made in moves during the argumentation stage of a dialogue. For example, if one party makes a particular assertion during the argumentation stage and the other party challenges that assertion, then the normal rule is that the party who made the assertion must supply some kind of support using an argument to back it up.

Hamblin tells us (1970, 274) that the concept of burden of proof is replaced in his system with the simpler concept of initiative, which appears to coincide with the concept of local burden of proof. The burden of proof rule in the dialogue system PPD is local because it applies during the sequence of moves in the argumentation stage where one party challenges some specific assertion made by the other party at a previous move. The concept of formulating a standpoint in a critical discussion is one of global burden of proof that applies over the whole sequence of dialogue from the opening stage to the closing stage.

There is a growing literature on burden of proof in argumentation (Kauffeld, 2003) and in work on formal dialogue models in artificial intelligence (Prakken, Reed and Walton, 2005; Gordon, Prakken and Walton, 2007; Prakken and Sartor, 2009). Importantly, this work has distinguished several types of burdens in persuasion dialogue as opposed to the widely accepted traditional assumption that there is a single concept of burden of proof. In legal argumentation in a trial there is a burden of persuasion set at the opening stage of the trial, and a burden of production of evidence is set as argumentative moves are made back and forth by the two sides during the argumentation stage. The burden of persuasion specifies which party has to prove some proposition that represents the ultimate claim to be proved in the case. The judge is supposed to instruct the jury on what proof standard has to be met. Whether this burden has been met or not is determined at the end of the trial. The burden of persuasion never shifts from the one side to the other during the whole proceedings. The burden of production specifies which party has to offer evidence on some specific issue that arises during the argumentation stage of the trial. According to recent work in artificial intelligence and law (Prakken and Sartor, 2009, 228), there is also a tactical burden of proof that is decided by the party putting forward an argument at some stage during the proceedings. The tactical burden is not ruled on or moderated by the judge. It pertains only to the two parties contesting on each side, enabling them to plan their argumentation strategies. The arguer must judge the risk of ultimately losing on the particular issue being discussed if he fails to put forward enough evidence to fulfill his tactical burden of proof. In legal argumentation, the burden of persuasion is a global burden of proof, whereas the burden of production and tactical burden are both local burdens of proof.

4. Situating Support Requests in Types of Dialogue

One can see from Hamblin's (1970, 256) distinction between formal and descriptive dialectic that he envisaged the advent of diverse formal dialogue models that can be applied to different kinds of discussion formats like those found in a legal trial or legislative debate. But he did not go so far as to make a systematic attempt to define or classify these different types as goal-directed structures. Since then, the literature has gone on to build formal models of different types of dialogue. As indicated in Chapter 1, a formal dialogue is defined as an ordered 3-tuple *O, A, C* where *O* is the opening stage, *A* is the argumentation stage and *C* is the closing stage (Gordon and Walton, 2009, 5). At the opening stage, the participants agree to take part in an identifiable type of dialogue that has a collective goal.

One might raise the objection here is that it is improper to speak about the collective goal of a dialogue type because neither dialogues nor dialogue types are sentient entities. Only participants may have goals, and it is improper to speak of the dialogues themselves as having goals. This point is disputable, but it is not at all obvious that only sentient beings can have goals. Activities can also have goals. Also, collective bodies, such as corporations or states, are not sentient beings (even though sentient beings belong to them) and can have goals over and above the individual goals of their members. For example, it is typical for organizations, like corporations for example, to formulate a "mission statement" that explicitly asserts what the founders or members have agreed upon to be the collective goal of the organization.

In formal dialogue systems the goal of the dialogue needs to be distinguished from the individual aims of the participants, and even from their shared purposes, in order to address the problem that in real conversations, some people engage in apparently purposeful interactions merely to distract or waste the time of the other participants. It is precisely for this reason that the goal of an activity needs to be distinguished from the individual aims of the participants. In a deliberation dialogue, for example, the goal of deliberation, namely reaching a decision on how to act, needs to be recognized, independently of whether any or all of the participants are seriously deliberating in order to fulfill the goal of reaching a rational collective decision on what to do.[1]

During the argumentation stage, the two parties, just as illustrated by the Hamblin Why-Because System with Questions, take turns making moves containing a speech act, like asking a question, making an assertion or putting forward an argument to support a claim. Just as in Hamblin's dialogues, when each party makes a move, statements are inserted into or retracted from his/her commitment store. Dialogue rules (called protocols in AI) define what types of speech acts are allowed, when each type of speech act is allowed as a move by a party and how each speech act made in a move can be replied to at the next move by the other party (Walton and Krabbe, 1995). The type of dialogue is determined by its initial situation, the collective goal of the dialogue shared by both participants and each individual participant's goal.

---

[1] I would like to thank Erik Krabbe for bringing out these helpful points in answer to my questions about dialogues having collective goals in an email dialogue on October 15, 2011.

The seven basic types of dialogue recognized in the argumentation literature were shown in Table 1.1. Persuasion dialogue is adversarial in that the goal of each party is to win over the other side by finding arguments that defeat its thesis or casts it into doubt. "Persuasion dialogue" has now become a technical term in artificial intelligence, and there are formal computational models of it (Prakken, 2006). Critical discussion (van Eemeren and Grootendorst, 1992) is classified (Walton and Krabbe, 1995) as a type of persuasion dialogue.

One needs to raise the question of what the rationale is for having a burden of proof in a persuasion dialogue. The aim of each party in a persuasion dialogue is to try to get the other party to make assertions, and then use these assertions as commitments to prove one's ultimate conclusion. The best defensive strategy is to make as few commitments as possible yourself, and the best offensive strategy is to try to get the other party to make as many commitments as possible. But once a proponent has made such a claim, and it has been challenged by the other side, it is generally in her interests to support it as strongly as possible with convincing arguments. Thus, there would seem to be no strategic reason to have to back up your assertion in a persuasion dialogue if you see the persuasion dialogue as a zero-sum game in which the goal of each party is to persuade the other, and the winner is the party who first accomplishes this aim. For example, in the critical discussion type of dialogue of van Eemeren and Grootendorst, each party has the ultimate goal of persuading the other to accept his or her thesis. The first party to do this wins, and the other party loses. The goal of resolving the conflict of opinions is accomplished when one party produces an argument that proves his or her thesis. In this type of dialogue both parties have plenty of incentive to support their assertions needed to prove their final thesis. No further incentive, in the form of a burden of proof rule, is needed.

For example, Hahn and Oaksford (2007, 47) agree that it makes sense to have a global burden of proof at the opening stage of a critical discussion, but they question why we need to have a local burden of proof for each individual claim in an argumentative exchange. In their opinion the risk of failing to persuade by not providing proof of some particular claim that has been questioned is a relatively small factor in the outcome of the dialogue. They see the local burden of proof as "entirely external to the dialogue and not a burden of proof in any conventional sense" (Hahn and Oaksford, 2007, 47). This questioning of what function burden of proof has in a persuasion dialogue is quite legitimate.

Inquiry is quite different from persuasion dialogue because it is cooperative in nature, unlike persuasion dialogue, which is much more adversarial. The goal of the inquiry is to prove that a statement designated at the opening stage as the hypothesis is true, using a high standard of proof. A central goal of inquiry is to prove a hypothesis to a sufficiently high standard so there will be no need to reopen the inquiry once it has been closed. Thus, meeting a burden of proof is fundamentally important in an inquiry.

Deliberation is also a collaborative type of dialogue in which parties collectively steer group actions toward a common goal by agreeing on a proposal that can solve a problem affecting all of the parties concerned while taking their interests into account (McBurney et al., 2007, 98). At the opening stage, the governing question cites a problem that needs to be solved cooperatively by the group taking part in the deliberation, a problem that concerns choice of actions by the group. During a later stage, proposals are put forward that offer answers to the governing question. The goal of the dialogue is not to prove or disprove anything, but to arrive at a decision on which is the best course of action to take.

Hamblin's remark (1971, 137) that his dialogue systems are "information-oriented" suggests that they should be classified as information-seeking dialogues where the collective goal of the dialogue is the exchange of information between the participants. But his discussions of rules for his Why-Because System with Questions strongly suggest a persuasion type of dialogue. A persuasion dialogue is one where the proponent has the goal of getting the respondent to commit to a thesis designated at the opening stage of the dialogue. She can only accomplish this goal by presenting an argument that fits a valid form of inference and has premises that the respondent is committed to. This aspect of persuasion dialogue is particularly strongly suggested by Hamblin's formulation of the CtP rule. If one party is going to justify a statement, surely she needs to use an argument with premises that are commitments of the other party. Otherwise the argument will not be useful to persuade rationally the speaker to come to accept the statement that needs justification. Persuasion, in this sense (referring to rational persuasion), refers to the effecting of a change in the respondent's commitment set (Walton, 1989). If the proponent can carry out this designated task, called the burden of persuasion by Prakken and Sartor (2009), she wins the dialogue as a whole. However, she typically has to use a lengthy chain of arguments to persuade the respondent one step at time, and the respondent has possibilities for retracting his commitments along the way.

5. Specifications for a Why-Because System with Questions

Hamblin's approach of discussing rules of dialogue in a flexible way, instead of going ahead to build precise systems with rigid rules, seems wise in retrospect. It is a precursor of the approach of Reed (2006), who has advocated assisting with the computational work of building a multiplicity of dialogue systems for many diverse applications in computing through what he calls a dialogue system specification (DSS). This approach provides a more convenient method for setting up formal dialogue systems that are useful for modeling argumentation. For our purposes we don't need to worry about resolution requests or yes-no questions, and we can work with an even simpler specification system that lacks these speech acts. We are primarily interested in burden of proof, so mainly we need to be concerned with support requests and assertions.

The problem taken up in this section is how to build a DSS that is an extension of Hamblin's system and that has capabilities for dealing with argumentation structures that were unknown in 1970. What is needed to cope with burden of proof is a support response mechanism that is more inclusive than the one considered by Hamblin. He used a deductive system of propositional calculus, or some comparable deductive system of classical logic, as his language for the Why-Because System with Questions. But at this point in the development of formal dialogue systems, it is necessary to take defeasible reasoning into account. The rake theft dialogue illustrates this need very well, for nearly all of the arguments put forward in it are defeasible. We need to allow a participant who responds to a request for support of a claim to use defeasible rules of inference as well as deductive rules of inference.

In this new system, support requests have to take a special form. There is only one rule of inference, *modus ponens* (MP), but it can take two forms, strict MP and defeasible MP. Strict MP, familiar in deductive logic, has a conditional premise that is not open to exceptions. Defeasible MP has a conditional premise that is open to exceptions (Verheij, 1999, 115; Walton, 2002, 43). The strict MP form of argument that we are familiar with in deductive logic has one premise that is a material conditional →. It has this form: $A \rightarrow B$; A; therefore B. Defeasible MP has the following form, where $A \Rightarrow B$ is the defeasible conditional: $A \Rightarrow B$; A; therefore B. For example, if something is a bird then generally, subject to exceptions, it flies; Tweety is a bird; therefore Tweety flies. This argument is the canonical example of defeasible reasoning used in computer science. If we find out that Tweety is a penguin, the original defeasible MP argument defaults. It is shown in (Bex, Prakken, Reed and Walton, 2003) how defeasible conditionals of these kinds can be treated as generalizations in legal reasoning, and the same point, applies in a case of ethical reasoning like the rake theft example.

For this new Why-Because System (WB System) we need to use a defeasible logic. Defeasible logic (Nute, 1994; 2001) is a rule-based nonmonotonic formal system that models reasoning used to derive plausible conclusions from partial and sometimes conflicting information. A conclusion derived using defeasible logic is subject to retraction if new information is presented that shows there is an exception to the general rule. The basic units of any system of defeasible logic are facts and rules. There are two kinds of rules: strict rules and defeasible rules. Facts are atomic statements that are accepted as true or not within the confines of a type of dialogue. To prove a conclusion using defeasible logic you have to carry out three steps (Governatori, 2008): (1) give arguments for the conclusion to be proved, (2) consider all the possible counterarguments that can be offered against the conclusion and (3) defeat these counterarguments by either showing that some premises in them do not hold or by producing stronger counterarguments against them. Defeasible logic moves forward in a dialectical fashion by bringing forward the pro and contra arguments relevant to a claim at issue. The conclusion at issue is proved if the arguments supporting it are stronger than the arguments against it. In the dialogue system ASD (Reed and Walton, 2007) defeasible argumentation schemes can be used as inference rules.

These considerations take us back to the support request speech act in the WB System formulated in Section 1. In this system, the speech act "Why A" is taken as a request for the addressee to supply an argument that would give the speaker a reason for him to accept A. What is requested is an argument with A as its conclusion and it needs to have one or more premises. In the WB System there are only two rules of inference that the addressee can use for this purpose, deductive MP and defeasible MP. This approach is broader than Hamblin's dialogue system, which had no provision for use of defeasible inference rules. One might ask whether other rules of inference can be added. For the present, there are controversies about which rules can be added. The current trend in applications of defeasible logic in artificial intelligence is to use defeasible MP, but not to use other forms of inference such as contraposition and *modus tollens* (Caminada, 2008, 111). Two systems of defeasible logic of this sort are (Reiter, 1980) and (Prakken and Sartor, 1997).

Hamblin's system has the rule that any assertion made by one party is open to challenge by the other party. This rule is appropriate for certain types of dialogue, like the Socratic style of dialogue, where all assumptions are subject to critical questioning. However, it has been emphasized by van Eemeren and Grootendorst (1992) that resolving a conflict of opinions by a critical discussion depends on both parties agreeing to common starting points. They agree at the opening stage not to dispute these propositions because challenging them during the argumentation stage would be a waste of resources. An example is the proposition 'Los Angeles is in California.' Continually challenging such propositions could well hinder the goal of resolving the issue at stake. A proposition accepted by both parties as common knowledge should not have to be proved, and cannot be disproved, at least within the confines of the critical discussion that is underway.

In law, as well, propositions that any reasonable person would agree there is no doubt about do not need to be proved. They are accepted by judicial notice. Propositions admitted into evidence in a trial need to be proved, but if every single assumption needed be proved, it might take years to solve even the simplest case. Judicial notice is a rule in the law of evidence that allows a proposition to be introduced as evidence in a trial if its truth is so well known that it is accepted as common knowledge.

Common sense systems in AI also contain many examples of common knowledge. The open mind common sense system (OMCS) includes such propositions as "If you hold a knife by its blade then it may cut you." and "People pay taxi drivers to drive them places." under the heading of common knowledge (Singh, Lin, Mueller, Lim, Perkins and Zhu, 2002, 3). Freeman (1995, 269) classified a proposition as a matter of common knowledge if many, most or all people accept it.

Hamblin (1970, 278) recognized the need to have "popular beliefs" in a dialogue system for representing debates and other real instances of argumentation. He proposed having a list of statements in the dialogue representing commonly accepted beliefs (278). Accordingly, in the new WB System, each participant has a subset of its commitment set called a "common knowledge set." This set contains propositions accepted as common knowledge by both parties at the opening stage of the dialogue. These commitments are different from the other commitments in a participant's commitment set because they cannot be retracted once each participant has agreed to them at the opening stage. Another feature is that when one of them is asserted, it does not have to be proved, and is even immune from challenge by the other party. Hence, there are limits on burden of proof in the WB System. A first party does not have a burden of proof to support his assertion with an argument when a second party challenges it if the proposition asserted is in the common knowledge commitment set in the dialogue. A stronger version of the WB System, which I call WB+, even has a rule forbidding such challenges. WB and WB+ are not presented as complete formal dialogue systems, but as dialogue system specifications following the style of Hamblin's discussion of rules summarized in Section 1.

6. Analysis of the Argumentation in the Example

A dialogue representing a very simple analysis of the arguments on both sides in the rake theft example is represented in Figure 5.1.

The premises and conclusions are shown as text boxes containing statements (propositions), and the arrows represent inferences from premises (or from a single premise) to a conclusion. The argumentation on the two sides is presented in a format of two columns, each representing the sequence of argumentation attributed to a particular participant. This initial analysis of the structure of the argument is meant to be only a very simple representation. Subsequently, a more refined analysis will be offered.

On the left we see Alfred's ultimate conclusion to be proved at the top, the statement that Bob stole Kathy's rake. In the right column at the top, we see Dana's ultimate conclusion, the statement that Bob did not steal Kathy's

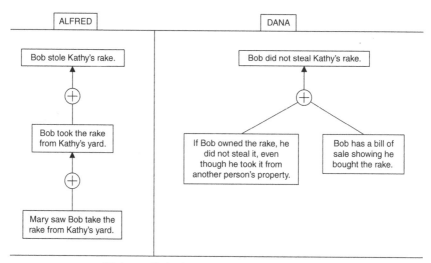

FIGURE 5.1. Dialogue-Style Argument Map of the Rake Theft Example

rake, the opposite (negation) of Alfred's conclusion to be proved. Each side has an argument to support its ultimate conclusion. Alfred brings forward the premise that Bob took the rake from Kathy's yard. This argument is clearly a defeasible one, and a fairly weak one at that, because there might be a variety of reasons why Bob took the rake without stealing it. He might have been simply borrowing it, for example, and have had Kathy's permission to take it from the yard. Following that, Alfred uses an argument from witness testimony to back up his premise that Bob took the rake from Kathy's yard, claiming that Mary saw Bob take the rake. In the right column, we see an argument with two premises. Using the standard argument mapping notation, it is represented as a linked argument in which the two premises go together to support the conclusion. Clearly the argument has a defeasible *modus ponens* structure, but we do not represent this feature anywhere on the argument map in Figure 5.1. It will be shown in the next argument maps.

What we do see from Figure 5.1 is its dialogue structure, showing that the argument has two sides. Each of the two parties has a thesis to be proved, and the thesis to be proved by the one side is the opposite of thesis to be proved by the other side. Each side proceeds to present arguments to support its thesis. It would appear from the dialogue classification typology presented earlier that this argument fits the structure of a persuasion dialogue. It is a dispute, a conflict of opinions in which each side has a thesis that is opposed to the thesis of the other side. Each side tries to present the most convincing arguments to show the other side that the first side's thesis is acceptable.

The argument map in Figure 5.2 shows a more detailed representation of the structure of the argument on one side in which some implicit

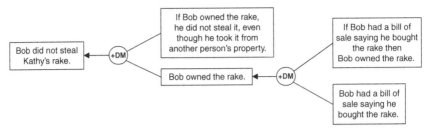

FIGURE 5.2. Map of Dana's Argument in the Rake Theft Example

premises are represented. This way of representing the argument reveals more of its structure, and in particular, it shows clearly how defeasible *modus ponens* (DM) is used to draw inferences from premises to a conclusion.

Once again the premises and conclusions are shown as statements in text boxes, but each argument itself is shown as a node that is intermediary between the premises and the conclusion. A convergent argument is displayed as two separate arguments, each with its own separate node. A linked argument is shown as an argument with more than one premise leading to the same node that leads to the conclusion. For example, in the top argument on the left, the conclusion is the statement that Bob did not steal Kathy's rake. It is a linked argument because we can see that its two premises both lead to the node containing DMP, which in turn leads to the conclusion.

In Figure 5.2, we can see that two of the premises are implicit premises that have been inserted into the argument based on an interpretation of how the sequence of reasoning should run. In this instance, both conditional premises are implicit. A representation of the other side of the argument is given in Figure 5.3.

Another aspect of the argumentation in the rake theft example is that Dana attacked Alfred's argument from witness testimony by arguing that Mary has lied in the past. This part of the argumentation is shown in Figure 5.4.

In Figure 5.4, the argument at the bottom, with the conclusion that there is a reason for doubting Mary's reliability as a witness, is a con argument against the argument to the right of it. One premise of that argument is no longer accepted.

Now each party has a good argument to support its contention that its thesis can be supported by evidence. But there is one other argument to be considered. Recall that at move 6, Alfred asked Dana to prove that Bob did not steal Kathy's rake, and Dana replied that the rake was not Kathy's property. To prove this claim he offered the argument that the rake was not Kathy's property, and supported it with the claim that Bob had a bill of sale showing he had bought the rake. This argument can be seen as a strong refutation, because it shows that given the premises, it is not possible that

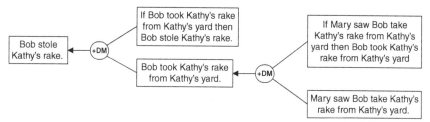

FIGURE 5.3. Map of Alfred's Argument in the Rake Theft Example

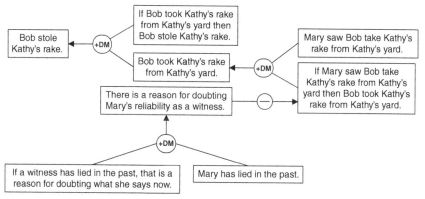

FIGURE 5.4. Dana's First Counterargument to Alfred's Argument

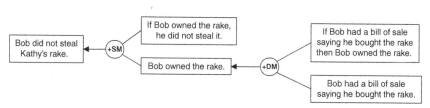

FIGURE 5.5. Dana's Second Counterargument to Alfred's Argument

Bob stole Kathy's rake. The reason is that it was not Kathy's rake. You can't steal something that is yours. This generalization is true by definition if 'theft' can be defined as stealing somebody else's property. As represented in Figure 5.5, this argument is based on the argumentation scheme for SMP, the deductive form of *modus ponens*.

Dana wins the argument because it is a necessary condition of stealing something that the object stolen was not the property of the person claimed to have stolen it. Unless Alfred can refute Dana's premise that Bob had a bill of sale saying he bought the rake, Alfred's claim that Bob stole Kathy's rake is strongly refuted.

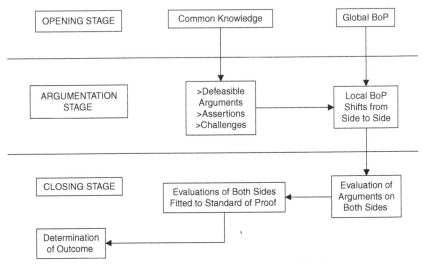

FIGURE 5.6. Outline of the Structure of the WB System

7. Solution to the Problem of Evasion and Shifting of Burden of Proof

There are two basic problems with burden of proof from the point of view of detecting argument abuses and unfair sophistical strategies (Ricco, 2011), evasion of burden of proof and shifting of burden of proof. However, whether such moves are reasonable or fallacious is to be determined in specific cases by examining the particulars of the case. The reason is that, as we have shown, in some cases, failing to give an argument to support one's claim when a why question is posed is a reasonable response in a dialogue. For example, if the proposition queried is an item of common knowledge, no argument supporting it is required to be furnished.

The analysis of the rake theft example shows that in any example of argumentation there is a sequence of arguments on each side of the dialogue. On White's side of the dialogue, there are arguments supporting White's claim and attacking Black's claim. On Black's side of the dialogue, there are arguments supporting Black's claim and attacking White's claim. The burden of proof set at the opening stage of the dialogue determines what thesis each party has to prove, and determines how the burden of proof is apportioned between them. It also sets the standard of proof in place that determines how strong a winning argument needs to be in order to prevail (Gordon and Walton, 2009). Once these elements are set in place, the argumentation stage runs through the speech acts put forward by both sides in their moves. Once the argumentation stage is finished and all the arguments are in, the closing stage is reached. At this stage, the burden of proof (BoP) set in place at the opening stage is used to determine which side had the winning argument, or whether there is no winner, if that is the outcome.

The arguments running through the argumentation stage can be represented as a large argument map in which all the premises and conclusions are connected to each other, and the argumentation schemes for the arguments that join them together in two specific arguments are labeled as SMP or DMP. In the analysis of the rake theft example, relatively specific arguments that display the chaining together of some premises and conclusions and larger chain of argumentation are represented in Figures 5.2–5.5. As described in Chapter 2, Prakken and Sartor (2009) have shown how a formal dialogue model can be used to evaluate such argumentation chains and judge the outcome in legal cases based on burden of proof. The basic outline of how such an argumentation system works is shown in Figure 5.6.

In a legal case, the burden of proof might be set higher on one side. For example, in a criminal case, the prosecution has to prove its claim beyond a reasonable doubt, whereas the defense wins if the prosecution's case is too weak to meet that standard of proof. Something like this setting of different burdens of proof for the ultimate claims of the two sides may also play a role in the argumentation in the rake theft example. Alfred has made a very serious allegation by claiming that Bob stole Kathy's rake and has therefore committed theft, an act that is (defeasibly) morally wrong. Dana has only made the claim that Bob did not commit theft, which seems like not such a serious allegation by comparison. Thus, ethically speaking, it might be fair to set a higher standard of proof for Alfred's argument to be successful in proving its claim. If so, Dana's argument shown in Figure 5.5 should prevail over Alfred's. It should be seen as a refutation of Alfred's argument, unless Alfred can introduce further evidence that would defeat it. Because Alfred has not done so at this point in the argument, Dana's side has the stronger argument and prevails in the dialogue. How much stronger it needs to be depends on the standard of proof set for it at the opening stage. But if that standard is taken to be preponderance of the evidence, meaning that the stronger argument prevails even if it is only slightly stronger (Gordon and Walton, 2009), then Dana wins.

The rake theft example showed that making a support request move could backfire in some instances. In this example, when Dana replied to Alfred's burden shifting question it provided an opportunity for Dana to produce an argument that proved to be so strong that it refuted Alfred's ultimate claim to be proved, thereby winning the dialogue for Dana. It shows that aggressively pursuing a sequence of why questions may not always be such a good idea in a persuasion dialogue.

What was shown is that to bring the Hamblin system up to date and deal with problems of burden of proof discussed since his time, several features have to be added in for this purpose. One of these features is the capability for dealing with defeasible arguments, because as the rake theft example showed, all the arguments had a DMP structure, except for one that had an SMP structure. The example shows how the WB system handles both kinds of *modus ponens* arguments, and so can better deal with issues of burden of proof. Hamblin did not attempt to deal specifically with burden of proof in his writings on formal dialogue systems and fallacies, preferring instead to use the simpler concept of initiative (Hamblin, 1970, 274) to represent the phenomenon of the shifting of the burden of proof during kinds of arguments where claims are made and need to be defended. By incorporating not only the capability for defeasible reasoning, but also the capability for modifying speech act rules in a way that can throw light on problems with burden of proof, the WB System reveals the power of the simple but flexible systems that Hamblin devised when it comes to analyzing fallacies and other highly significant phenomena of argumentation like the shifting of burden of proof. Building formal dialogue systems to provide models to represent the science of reasoning underlying the use of the notion of burden of proof as a way to analyze and evaluate argumentation has proved to be an extremely useful way of solving logical problems, like those represented by the fallacy of argument from ignorance.

Another feature of the WB System is the capability for managing arguments that depend on common knowledge. It is important to see that not all propositions have to be supported with arguments when challenged by the other side using a why request. If the proposition represents common knowledge, there is no obligation on the respondent to make any attempt to prove it. Also, if the proposition queried by the speaker is not a commitment of the hearer, the hearer does not have to respond to a why question by providing justification for it. Allowing for this possibility is accommodated by Hamblin's three responses rule, and no doubt, it was his awareness of the complications that can arise from matters of burden of proof that led him to formulate his rule in the way he did.

The rest of this chapter presents a simple formal dialogue system CBCK that fits the dialogue system specifications for WB+. It is argued that burden of proof needs to be defined, utilized and set in different ways in different types of types of dialogue. However, by concentrating on a very simple system, a basic outline is given to show how burden of proof can be defined, set and utilized in any of these types of dialogue by extending Hamblin's insights about dialogue systems. The system CBCK is built onto the more basic system CB by adding a set of propositions called a common knowledge set to the commitment sets of both participants at the opening stage of the dialogue.

8. Speech Acts in Dialogue Systems

Dialogue models of argumentation of the kind developed in (Walton and Krabbe, 1995) are now proving their worth as tools useful for solving many problems in argumentation studies, artificial intelligence and multiagent

TABLE 5.2. *Six Fundamental Types of Speech Acts*

Speech Act	Dialogue Form	Function
Question (yes-no type)	S?	Speaker asks whether S is the case.
Assertion (claim)	Assert S	Speaker asserts that S is the case.
Concession (acceptance)	Accept S	Speaker incurs commitment to S.
Retraction (withdrawal)	No commitment S	Speaker removes commitment to S.
Challenge (demand for proof of claim)	Why S?	Speaker requests that hearer give an argument to support S.
Put Argument Forward	$P_1, P_2, ..., P_n$ therefore S.	$P_1, P_2, ..., P_n$ is a set of premises that give a reason to support S.

systems. Many formal dialogue systems have been built (Bench-Capon, 2003; Prakken, 2005, 2006), and through their applications (Verheij, 2003), we are getting a much better idea of the general requirements for such systems, and how to build them. Reed (2006) has provided a dialogue system specification that enables anyone to construct a formal dialogue model of argumentation by specifying its components and how they are combined (Reed, 2006, 26). This dialogue specification system provides a more convenient method for setting up formal dialogue systems of kinds that are useful for modeling argumentation technologies in computing that have been built and are currently being built for various applications. According to the argument of this chapter, a variant on Reed's dialogue system specification can also be applied to the project of building a dialogue system specification for burden of proof.

There can be many different types of speech acts in different types of dialogue. In a persuasion dialogue, the speech act of putting forward an argument is very important, while in a deliberation dialogue, the speech act of making a proposal is very important (Walton, 2006). A good place to begin is with the most common and fundamental types of speech acts in persuasion dialogue as identified in Table 5.2.

A statement is taken to be equivalent to a proposition. An assertion is different from a statement. A statement is an entity that is asserted in an affirmative sentence in a language, and can be true or false. An assertion is a speech act made in a dialogue. It contains a statement and makes the claim that this statement is true. Commitment is very close to the meaning of acceptance, but is not exactly the same thing. Each participant in a dialogue has a commitment set (commitment store), and at each move statements (propositions) are inserted into a participant's commitment set or removed from it, depending on the move made and commitment rules for the dialogue. How commitment should be incurred or retracted (Walton and Krabbe, 1995,

123) is a function of five factors: (1) the kind of speech act made at a move, (2) the type of dialogue in which the move was made, (3) the goal of the dialogue, (4) the speaker's role in the dialogue and (5) the rules for that type of dialogue (including, of course, the commitment rules).

The three speech acts that are most central to formal argumentation dialogue structures of the kind that have been investigated so far are those of making an assertion, putting forward an argument and challenging an assertion or argument. Making an assertion is to make a claim that some proposition is true, or at any rate acceptable. An assertion is challenged by requesting that the proponent prove it. This move means essentially that the proponent has to give some argument to support the claim.

There are a wide variety of views in the philosophy literature on the nature of assertion. Rescorla (2009b, 99) defines these views into two main models that have been put forward – the restrictive model of assertion and the dialectical model of assertion. According to the restrictive model, a norm restricts the propositions one may assert. There are four main candidates for such a restrictive norm advocated by philosophers. One should assert only propositions that one knows to be true. One should assert only propositions that one believes. One should assert only true propositions. One should assert only propositions that one believes with sufficient warrant. The dialectical model treats assertion as constituted by its role in a dialogue exchange in which two parties ask for and give reasons. According to the dialectical model, there are no constraints on making an assertion. The dialogue norms only constrain how a participant must react if the other party challenges his assertion.

The motivating idea behind the dialectical model of assertion derives from the work of Hamblin (1970, 1971) on formal dialogue systems. Hamblin's abstract formal models of dialogue were meant to be normative, in the sense that they model what moves a rational participant must make, following the dialogue rules, like the commitment rules (Mackenzie, 1981). Hamblin introduced the notion of a commitment store in formal models of dialogue as his central concept for the analysis of argumentation. When a participant in a dialogue makes a commitment to a proposition, the proposition is then inserted into his commitment store. In his view, the main defining feature of the move in dialogue of making an assertion is that it commits the party making the assertion to the proposition that was asserted. This close connection between assertion and commitment is brought out very well by a remark of Searle (2001, 184): "There is no gap at all between a statement and committing oneself to its truth."

The defining characteristic of assertion has a normative ingredient, meaning that when a party makes a commitment, he or she is then bound to it. This was clearly articulated by Peirce, "This ingredient, the assuming of responsibility, which is so prominent in solemn assertion, must be present in every genuine assertion" (1934a, 386). This remark brings out the idea that when you make an assertion, that speech act or move binds you to what you asserted, until such time you retract commitment to it, if you can.

Putting forward an argument is a means of trying to get the respondent to accept that argument. More specifically, it is an attempt to get the respondent to accept the conclusion of the argument, given that the premises are acceptable to him. However, the issue is complicated by the fact that we sometimes put forward arguments hypothetically, not to get the respondent to accept the conclusion, but just to see where the argument might lead. Ruling this possibility out for the moment, however, the main use of putting forward an argument is to get the respondent to accept its conclusion. The question of how an argument should be challenged is even more complex, and subject to different views. It is often said that there are only three ways to challenge (attack, critically question, rebut, refute) an argument, by questioning a premise, by putting forward a counterargument that has the opposite conclusion of the given argument or by attacking the inferential link between the premises and the conclusion. The third way is sometimes called "undercutting the argument," as opposed to the second way that is called "rebutting the argument," following Pollock's distinction between rebutters and undercutters.

9. The Dialogue Systems CB1 and CB1CK

The term "acceptance" is very close to the meaning of commitment, but here we use it in a more precise way. Acceptance is a speech act whereby a participant directly takes on commitment to a statement S by saying "I accept S." Sometimes this expression only means "I concede S," but a distinction needs to be drawn between commitment and mere concession. The distinction between assertion and concession is not significant in all types of dialogue, but sometimes it can be. A participant in a dialogue is committed to his assertions in a different way from being committed to his concessions (Walton and Krabbe, 1995, 134). In some cases, a participant can concede a statement S for the sake of argument without being committed to defending it if challenged. However, when a participant asserts S, that means he is claiming that S is true, and therefore he is obliged to defend S if challenged to do so by the other party. This obligation is called the burden of proof. Commitment to concessions, in contrast, does not have a burden of proof attached (Walton, and Krabbe, 1995, 134). In some instances, saying you accept a statement only indicates concession of the statement accepted, while in other instances acceptance implies a commitment to defend the statement if it is challenged.

In some systems, distinctions are also drawn between other kinds of commitments. For example, the formal persuasion dialogue system PPD_0, distinguishes between explicit commitments and implicit commitments that are derived by implication from explicit commitments (Walton and Krabbe, 1995, 149–154).

To illustrate very simply how the foregoing notions can be put together in a formal dialogue system of the Hamblin type, we use the dialogue system CB1, based on the dialogue system CB[2] from (Walton, 1984, 133–135). CB1 has four locution rules, five commitment rules and three dialogue rules (Walton 1984, 133–135). To formulate these rules we need to start with the following definitions of immediate consequence and consequence from (Walton, 1984, 132–133). A statement T is an *immediate consequence* of a set of statements S_0, $S_1,\ldots, S_n$ if and only if '$S_0, S_1,\ldots, S_n$ therefore T' is a substitution instance of a rule of inference accepted in the dialogue. A statement T is a *consequence* of a set of statements $S_0, S_1,\ldots, S_n$ if and only if T is derived by a finite number of immediate consequence steps from immediate consequences of $S_0, S_1,\ldots, S_n$. To implement these two definitions some rules of reasoning for CB1 need to be adopted. For example, *modus ponens* could be adopted as the only rule. Alternatively, other rules for deductive reasoning could be adopted, as in CB. Or for that matter, some defeasible argumentation schemes could be adopted as rules of inference, as in the dialogue system ASD (Reed and Walton, 2007). CB1 is minimal, and highly flexible in this respect. It is a basic system that can provide a platform for building more complex systems like PPD_0.

The names for the four types of procedural rules (protocols) presented as follows are expressed in more or less the Hamblin style. In newer systems, other terms may be used. For example, what are called the dialogue rules are now often called the pre- and postcondition rules. What are called termination rules are called strategic rules by Walton and Krabbe.

Locution Rules

(1) Assertions: Statement letters, S, T, U, ..., are permissible locutions, and truth-functional compounds of statement letters.
(2) Withdrawals: 'No commitment S' is the locution for withdrawal (retraction) of a statement.
(3) Questions: The question 'S?' asks 'Is it the case that S is true?'
(4) Challenges: The challenge 'Why S?' requests some statement that can serve as a basis in proof for S.

Commitment Rules

(1) After a participant makes a statement, S, it is included in his commitment store.
(2) After the withdrawal of S, the statement S is deleted from the speaker's commitment store.

---

[2] The names of the dialogue systems started with the letters A, AA, AB, B, BA and so forth. CB was the third of the C systems. The idea was to start with the simplest systems and work up to more complex ones.

(3) 'Why S?' places S in the hearer's commitment store unless it is already there or unless the hearer immediately retracts his commitment to S.

(4) Every statement that is shown by the speaker to be an immediate consequence of statements that are commitments of the hearer then becomes a commitment of the hearer's and is included in his commitment store.

(5) No commitment may be withdrawn by the hearer that is shown by the speaker to be an immediate consequence of statements that are previous commitments of the hearer.

Dialogue Rules

(1) Each participant takes his turn to move by advancing one locution at each turn. A no-commitment locution, however, may accompany a why-locution as one turn.

(2) A question 'S?' must be followed by (1) a statement S, (2) a statement 'Not–S,' or (3) 'No commitment S.'

(3) 'Why S?' must be followed by (1) 'No commitment S' or (2) some statement T, where S is a consequence of T.

Termination Rules

(1) Both participants agree in advance that the dialogue will terminate after some finite number of moves.

(2) The first participant to show that his own thesis is a consequence derived only from premises that are commitments of the other participant wins the dialogue.

(3) If nobody wins as in (2) by the agreed termination point (some number of moves determined at the opening stage), the dialogue is declared a draw.

It should be noted that the rules of CB1 have been modified from the original version of CB from (Walton, 1984, 132–33). The most significant change is that Termination Rule 2 is no longer restricted to immediate consequences.

The obligation associated with burden of proof is expressed by dialogue rule 3. When a party is challenged by virtue of the other party asking "Why S?", he must immediately (at his next move) either (a) withdraw commitment to S or (b) produce an argument that fits one of the rules of inference, or an argument composed of a chain of such inferences, that has S as the conclusion.

It should be noted that both parties are free at any move to make any assertion they care to. The only exceptions stem from postcondition rules, and the rule expressing burden of proof in CB1 can be used as an example. If one party asks "Why S?," the other party cannot at his next move make an assertion, like asserting some proposition T for example. The reason is simply that the replying party is only allowed to withdraw a commitment to S or make an argument for S at this responding move. He cannot make any assertion he likes, or indeed, cannot make any assertion at all. However, generally, aside from such exceptions, a party in CB1 can make any assertion she likes at any move. However, participants in CB1 are not likely to make assertions unless compelled to, for strategic reasons. The goal of each party is to win the dialogue by finding an argument for his/her own conclusion based on the other party's commitments. Hence, making commitments that one is not compelled to make is not, in general, strategically prudent. The idea is to get the other party committed to any propositions that might be useful as premises to prove one's own conclusion. We can sum up by saying that there are no restrictions on making assertions in CB1, except for instances where making an assertion is not an appropriate move because of the turn-taking rules, or where it would not be strategically prudent to make an assertion. In the latter case, however, a party can make the assertion if he likes. It's just that he might not see it as strategically useful.

We can now use the simple system CB1 to build the more complex system CB1CK. CB1 has the property that any assertion made by one party is open to challenge by the other party. This property may be suitable for certain types of discussion, like a Socratic style of dialogue where all assumptions are subject to questioning and probing scrutiny. However, in many types of debates and everyday conversational arguments, this degree of freedom to challenge may be counterproductive. It has been emphasized in argumentation studies that resolving a conflict of opinions by reasoned argumentation depends on common starting points that both parties in a discussion share, and agree at the opening stage not to dispute (van Eemeren and Grootendorst, 1992). Challenging such a proposition during the argumentation stage would be a waste of resources, and would not help to move the discussion forward toward its goal of resolving the issue at stake. A proposition accepted by both parties as common knowledge, for example, should not be challenged and should have to be proved. An example would be the proposition 'Los Angeles is in California.' Freeman (1995, 269) classified a proposition as a matter of common knowledge if many, most or all people accept it. The open mind common sense system (OMCS)[3] included such propositions as 'If you hold a knife by its blade then it may cut you.' and 'People pay taxi drivers to drive them places.' under the heading of common knowledge (Singh, Lin, Mueller, Lim, Perkins and Zhu, 2002, 3). Such statements are defeasible generalizations that can be assumed to hold generally, *ceteris paribus*, but can be

---

[3] http://commonsense.media.mit.edu/cgi-bin/search.cgi.

defeated by exceptions to the rule. The kind of common knowledge that is very important in AI and cognitive science is based on ordinary ways of doing things familiar to all of us in everyday life. According to Schank and Abelson (1977), this kind of common knowledge is based on what they call a *script*, a body of knowledge shared by language users concerning what typically happens in certain kinds of stereotypical situations, and that enables a language user to fill in gaps in inferences not explicitly stated in a text of discourse.

To accommodate these views about the usefulness of having practical limits on the right to challenge, we introduce a second system called CB1CK (for common knowledge) constructed by extending CB1. In a CB1CK dialogue, each participant has a subset of its commitment set called a common knowledge set. This set contains propositions accepted as common knowledge by both parties at the opening stage of the dialogue. These commitments are different from the other commitments in a participant's commitment set because they cannot be retracted once each participant has agreed to them at the opening stage. Another feature that they have is that when one of them is asserted, it is immune from challenge by the other party. A complication is that they can be challenged in unusual circumstances, but in such a case, when a challenge is arguably legitimate, the issue of whether the challenged proposition needs to be defended is resolved by a shift to a metadialogue (Krabbe, 2003). In other words, challenging a common knowledge commitment is not impossible, but it is unusual, and it is harder to carry out than challenging an ordinary commitment. Without bringing metadialogue rules into play, we can say that in CB1CK there are limits on the right of a party to challenge.

What this means is that in CB1CK there should be limits on burden of proof. A party should not have a burden of proof to back up his assertion with a supporting argument when the other party challenges it, provided the proposition asserted is contained in the common knowledge set in the dialogue. Indeed, such a need to fulfill the requirement of burden of proof would never arise in such a case provided CB1CK has a rule forbidding such challenges. This means that for purposes of CB1CK, there are two choices for modifying the dialogue rules. One option is that the rule 3 in CB1 can be modified as follows, and an additional rule can be added to it.

(3*) 'Why S?' must be followed by (1) 'No commitment S' or (2) some statement T, where S is a consequence of T, unless S is a common knowledge proposition.

(4) If 'Why S?' is asked, and S is a common knowledge proposition, the hearer should reply that S is common knowledge and does not need to be proved.

The other option is to add a dialogue rule barring the making of challenges to a common knowledge proposition. Such a rule can be formulated as follows.

(4*) A challenge move 'Why *S*?' can only be made if *S* is not in the common knowledge commitment set.

Either adding the combination of 3* and 4 or adding the single rule 4* have the same effect as far as regulating burden of proof is concerned. It does not matter which option is chosen for our purposes here. However, it is simpler to choose 4 to add to CB1CK, and so that is what we will opt for.

The why question in CB1, following the Hamblin style, is basically a request for justification. When a proponent poses the question "Why *S*?," she is requesting that the respondent provide an argument that would compel her by logical reasoning to become committed to *S*. This way of configuring the part of the dialogue that pertains to burden of proof seems simple enough, but Hamblin (1970, 271) goes on to consider some additional rules that could be imposed on the asking and answering of why questions. The first rule postulates a precondition for the asking of a why question.

(R_0) Why Question Asking Rule: 'Why *S*' may not be asked unless *S* is a commitment of the hearer and not of the speaker.

Hamblin (1970, 271) justified R_0 with the comment, "Otherwise the 'Why' is academic." This justification is reasonable, in many instances, for a speaker, we can presume, would normally only have a reason to ask a why question concerning a statement the hearer is already committed to and the speaker is not. If the hearer is not committed to *S*, it would be inappropriate, perhaps even fallacious, for him to be forced to prove it by providing an argument for it at the next move. This move would now seem to be some kind of inappropriate shift of the burden of proof. If the speaker is already committed to *S*, it makes no sense for her to demand that the hearer prove *S*. What's the point?

The second rule postulates a postcondition for responding to a why question.

(R_1) Why Question Answering Rule: The answer to 'Why *S*,' if it is not 'Statement not *S*' or 'No commitment *S*,' must be in terms of statements that are already commitments of both speaker and hearer.

R_1 also appears to be a good rule for the kind of dialogue Hamblin has in mind, which roughly appears to be a persuasion type of dialogue. But it is difficult to judge from Hamblin's remarks whether he had persuasion dialogue in mind when he formulated this rule, or some other kind of dialogue.

A persuasion dialogue is one where a proponent tries to get a respondent to come to accept a thesis designated at the opening stage of the dialogue. She needs to do this by presenting an argument that fits a valid form of inference and has only premises that the respondent is committed to. In the system ASD (Reed and Walton, 2007) defeasible argumentation schemes also can be used as inference rules). Persuasion, in this technical and normative sense of the term (meaning rational persuasion), refers to the effecting of a change in the respondent's commitment set (Walton, 1989). If the proponent can carry out this designated task, which represents the ultimate burden of proof in the dialogue, she wins the dialogue as a whole. However, she typically has to use a lengthy chain of arguments to persuade the respondent one step at time, and the respondent has possibilities for retracting his commitments along the way. The respondent is successful in the dialogue if he can resist this process of rational persuasion. CB represents only a very simple and basic form of persuasion dialogue, and it has many technical problems in it that make it highly doubtful whether it represents a practical system of persuasion dialogue that would be applicable to real cases. However, there are much more refined systems (Walton and Krabbe, 1995; Prakken, 2006) that that are more complex, but that present solutions to some of these problems. Persuasion dialogue can be contrasted with other types of dialogue where the goal is to negotiate, deliberate about a choice or exchange information, as will be shown in more detail in Chapter 7.

10. Dialogue Systems with Argument and Explanation

In Chapter 2 it was stated that research on burden of proof is in a bad place because of the evidence showing that the beyond reasonable doubt standard, the tool that judges and juries have to use in order to decide whether a defendant is guilty of a criminal offense or not, cannot be formulated by means of a clear or precise criterion. Hence, we appear to be left without a method for evaluating the evidential reasoning on both sides of a disputed case to determine which side has the winning argument. In Chapter 2 we posed the pointed question of how artificial intelligence can move forward to define burden of proof, working from this very difficult position. Is there some way of modeling the standards of proof that can provide a way to move forward in light of the difficulties of plugging in numerical values for the thresholds informal system itself? Is there a way of implementing the standards of proof in computational models of dialogue suitable for modeling real cases where burden of proof is a central issue or problem in reconstructing and analyzing the evidential reasoning in the case?

Laudan (2006, 82) suggested that there is one possible route forward based on a proposal made by Allen (1993, 436), who proposed the following criterion, "if the prosecutor's story about the crime is plausible and you can conceive of no plausible story that leaves the defendant innocent, then convict. Otherwise, acquit." This proposal goes in a different direction from the usual one of trying to apply Bayesian calculations to probabilistic arguments. Instead, it is based on the notion that the two sides in the trial are each trying to present an explanation of what happened in the case from which one can infer that the accused party was guilty or not.

Pardo and Allen (2007, 224–225) suggest that writing on principles of proof in the law of evidence has focused primarily on probability theory of the Bayesian sort, and that the comparative neglect of explanation-based reasoning has been a mistake. Allen and Pardo (2007, 109–120) surveyed the literature on the many attempts to apply Bayesian probability theory to judicial proof, concluding that there is a deep conceptual problem with this endeavor. They have proposed an explanation-based model of standards of proof as a way to move forward. Using this model, they reformulate the standards of proof used in law as follows. The standard in civil law is that of preponderance of the evidence. According to Pardo and Allen (2007, 237), the question of whether this standard is met or not in any given case should be made by choosing between competing explanations. The better of the two explanations should be taken to be the argument that meets the standard of preponderance of the evidence, or if there are several explanations, the best one is the one that meets this standard of proof.

In criminal cases, where the standard is that of beyond a reasonable doubt, the defendant should be convicted if there is a sufficiently plausible explanation of the evidence showing guilt, and there is no plausible explanation consistent with innocence (Pardo and Allen 2007, 238–239). To meet the clear and convincing evidence standard, a party has to present an explanation that is sufficiently more plausible than the explanation given by the other side, and this explanation must be clearly and convincingly more plausible than that given by the other side. Pardo and Allen admit that these standards are vague, in that they leave open precisely how sufficiently plausible an explanation must be in order to meet them. No precise probability value is given. In response to this criticism they suggest that this admitted lack of precision may be a critique of the standards, but it is not a critique of the explanation-based model of the standards, because they are no more imprecise than attempts to implement them with numerical probability values.

Clearly the speech act of assertion is vital to understanding the nature of burden of proof. There appears to be broad agreement both with respect to the argumentation in law and in everyday conversational exchanges, that when two parties are engaged in argumentation, a burden of proof is incurred when one of them puts forward an assertion of a specific proposition, thereby making a claim that this proposition is true, and from our point of view of rational discourse, taking on an obligation to prove that it is true if called on to do so by the other party. As shown earlier, asking a why question is to make a request for an argument to support a claim. But in everyday conversational discourse, asking a why question can be ambiguous. There is also a dialogue system (Walton, 2011) for explanation in which one party requests explanation by asking a why question, and the other party responds with a speech act that provides an explanation that attempts to help the questioner to understand what he asked about. In this system, the function of an explanation is to transfer understanding from an explainer to an explainee in a dialogue. In this system, the speech act of asking of a why question is defined as a request by one party to the other to offer an explanation of something questioned by the first party. In the explanation dialogue system, there are precondition rules that set the conditions under which a party is allowed to request or supply an explanation.

The dialogue model of explanation utilizes a notion of understanding deriving from the companion notion of a script in cognitive science. A script is a connected sequence of events or actions that both parties understand using their common knowledge about the ways things generally can be expected to go in situations both are familiar with (Schank, 1986; Schank and Abelson, 1977; Schank and Riesback, 1981; Schank, Kass and Riesback, 1994). Speech acts of offering and accepting an explanation in a dialogue after the manner of (Walton 2011) can then be given.

- The explainer and the explainee are taking part in a dialogue in which they share common knowledge about some domain.
- The explainee asks a question requesting that explainer help her to understand what she assumes that the speaker understands.
- The explainer replies by offering an explanation using a script they can both understand based on their shared common knowledge.
- The explanation is successful if it helps the explainee to understand sufficiently what she asked about.
- If the explanation is successful the dialogue stops.
- If the explanation is not successful the dialogue can continue until repeated dialogue moves help the explainee to understand sufficiently what she asked about.
- The dialogue may have to be closed for practical reasons (lack of time or resources) even if it is not successful.

There are many details of how to extend the dialogue system described in Sections 1–9 to a system that has rules for explanation speech acts and other rules that can extend the system to explanations as well as arguments. However, there is recent research moving in this direction that will be explained in Chapter 6.

6

Solving the Problems of Burden of Proof

Erik Krabbe's pioneering article on metadialogues (dialogues about dialogues) opened up an important new avenue of research in the field, largely unexplored up to that point. His modest conclusion was that it was too early for conclusions (Krabbe, 2003, 89). Even so, by posing a number of problems along with tentative solutions, his article was a very important advance in the field. Hamblin (1970) was the first to suggest the usefulness of metadialogues in the study of fallacies. He proposed (1970, 283–284) that disputes that can arise about allegations that the fallacy of equivocation has been committed could be resolved by redirecting the dispute to a procedural level. This procedural level would correspond to what Krabbe calls a metadialogue (Krabbe, 2003). Other writers on argumentation (Mackenzie, 1979, 1981; Finocchiaro, 1980, 2005; van Eemeren and Grootendorst, 1992), as noted by Krabbe (2003, 86–87) have tacitly recognized the need to move to a metalevel dialogue framework, but none provided a metadialogue system. The study of metadialogues is turning out to be very important in argumentation theory and in computer science (Wooldridge, McBurney and Parsons, 2005).

In Chapter 6, it is shown how analyzing disputes about burden of proof is an important research topic for investigation in the field of metadialogue theory. It has recently been shown (Prakken, Reed and Walton, 2005) that legal disputes about burden of proof can be formally modeled by using the device of a formal dialogue protocol for embedding a metadialogue about the issue of burden of proof into an ongoing dialogue about some prior issue. In Chapter 8, a general solution to the problem of how to analyze burden of proof is yielded by building on this framework, using three key examples from Chapter 1 to show how disputes about burden of proof can arise. These three examples were presented in Chapter 1 as classic cases of burden of proof disputes, and now in Chapter 6 it is shown how current tools from argumentation theory and artificial intelligence based on metadialogues can be applied to the problems they pose.

In Section 10, some resources are set out for moving forward by combining argument and explanation to confront the impressive difficulties of building burdens of proof on precise computational standards of proof, especially the standard of beyond a reasonable doubt.

1. Problems To Be Solved

We now turn to a discussion of the problem cases that were presented in Chapter 1. The first example is the debate from the Canadian House of Commons in which the questioner requested that the government minister prove that depleted uranium is not being used for other than peaceful purposes. The minister replied that he was satisfied, on the basis of the available information, that the treaty is being respected. This reply is essentially an argument from ignorance. The minister replied that he had looked for any weaknesses in the treaty and found none. From this premise he drew the conclusion that there are none, and inferred that the treaty is being respected. He even tried to use more aggressive burden-shifting strategic maneuvering by asking the questioner to come forward with allegations of a more specific kind. However, at that point in the dialogue, another opposition member shouted that he should do a proper investigation, thereby attempting to shift the burden of proof back to the government representative to collect more evidence. Thus, here we have a classic case of a dispute about burden of proof.

The problem in this case is that the rules for Canadian parliamentary debate do not define burden of proof precisely, and it is left up to the speaker of the House to rule on such a dispute. What tends to happen in such a case is that the argument passes on to some different issue, and the speaker does not intervene on the burden of proof dispute one way or the other. The problem is left up to the public, who are presumably watching the debate, to determine which side has the better argument, by judging on which side the burden of proof lies. In analyzing and evaluating such a case, however, the normative framework of burden of proof set out in this chapter above can be applied, showing how the burden of proof shifts back and forth during these strategic maneuvers. Still, the problem is that because specific burden of proof requirements were not made at the opening stage, precise determinations about the burden of proof at the argumentation and closing stages cannot be made. If they had been made, however, they could be used, by also using a metadialogue if necessary, to solve the impasse. Here we have a problem about burden of proof, because the dispute cannot be resolved, and is simply left hanging. It is up to the audience to decide which side won the exchange.

The second case is an easier one in which a clear decision was arrived at by the U.S. Supreme Court. It was pointed out by the Supreme Court that the disabilities education act made no statement about the allocation of the burden of proof, and therefore that the normal default rule applied. This means that the ruling automatically followed from the opening stage that the parents had the burden of proof. The parents failed to use the potentially strong argument that the school districts have a natural advantage and expertise. But even so, the Supreme Court argued that this exception did not apply in this case because Congress had already obliged schools to share information with parents. The parents put forward the argument that putting the burden of proof on school districts will help to ensure that children receive a free special education for each disabled child. However, the Supreme Court, on examining the evidence from the argumentation stage, including the previous trials where this issue had been disputed, concluded that the argument did not provide sufficient grounds for departing from the normal default rule on burden of proof. On our theory, the Supreme Court was looking over all the argumentation that was offered by both sides in the previous trials and using that evidence to arrive at a conclusion about burden of proof, based on the requirements of burden of proof set at the opening stage. Unlike the case of a political debate, in a legal trial there are rules that set specific burdens of proof for each type of case that can be used to bring the argument to the closing stage with a definitive ruling on which party met its burden of proof. It is, in fact, a central feature of our system of law that it allows the trial to function as a device for resolving conflicts of opinions without the possibility of a stalemate arising.

The hardest case is the third one of the Dutch Supreme Court trial concerning the Los Gatos band. The analysis of this case requires the embedding of a metadialogue into the argumentation in the original ground-level argument. The decision of the Dutch Supreme Court was that Holland America had the burden of proof because they had made it impossible for Los Gatos to explain their reasons for not wanting to play.

2. Meta-Arguments and Metadialogues in Logic and AI

A dialogue, as we have defined it so far, is a framework of argumentation use in which participants in some definable type of conversation tacitly or explicitly make agreements about the rules of conduct they will observe. For example, participants might agree to take part in a critical discussion, in which the goal is that of resolving the conflict of opinions by rational argumentation. If this type of dialogue can be called a ground level dialogue (Krabbe, 2003, 83) a metadialogue can be called a dialogue about the dialogue, or about some dialogues. For example, there might be disagreement about the correctness of some moves in a dialogue. Participants may then move to a metadialogue in order to have a secondary dialogue on whether the move in the first dialogue can be judged to be correct or not by some criteria (Hamblin, 1970; Krabbe, 2003).

One subject that has been studied very little in formal dialogue theory is the dialectical shift, or change from one type of dialogue to another during a sequence of argumentation. Dialectical shifts were studied in (Walton and Krabbe, 1995), but very little systematic formal work seems to have been done on them since that time. Shifts are very important for formal dialogue theory as a tool for analyzing argumentation because they are very common in argumentation, because they are often associated with fallacies (Walton and Krabbe, 1995; van Laar, 2003) and because we can scarcely understand many argumentation phenomena without realizing that a shift underlies the argument. Some shifts are so common, and effect such a natural and smooth transition between two types of dialogue, that we need little in the way of technical tools to understand and manage the shift. Others are highly problematic, and some very special mechanism needs to be inserted between the two dialogues so that the shift can be procedurally managed in a proper and coherent manner that can be fair to the participants, and that can help us to analyze the argument that took place in an equitable and logically justifiable manner. One type of shift that can sometimes be highly problematic in this way is the burden of proof shift. As shown by the earlier examples, some such shifts are easy cases that can be managed without undue effort by simple rules that can be more or less automatically applied to the case. Others are hard cases that require systematic intervention, and cannot be fairly adjudicated without the intervention of a third party at a metadialogue level.

Wooldridge, McBurney and Parsons (2005), noting the steadily increasing attention to argumentation and informal logic given by the multiagent systems community over the past decade, argue that the formalization of such argument systems is a necessary step for their successful deployment. On their view, argumentation in dialogue is inherently metalogical, meaning that it does not just involve the asserting of statements about some domain of discourse, and putting forward arguments based on these statements, but also the making of arguments about these arguments at a higher level. To help assist the formal development of this notion, they define three tiers of a hierarchical argument system. The first level corresponds to statements about a domain. The second level defines the notion of an argument and captures notions like attack and defeat of an argument. The third level encompasses the process of reasoning about the arguments that were used at the second level. Wooldridge, McBurney and Parsons (2005, 7), note that metalogical systems have been widely studied in the past four decades in AI and logic, and little research has addressed the issue of meta-argument, an exception being Brewka (2001). One of their main aims is to put this idea of meta-argument on the map of argumentation research.

The object level does not contain arguments at all. It consists only of statements about a particular domain of discourse, and defines relations on the entities in this domain. In a legal setting, the object level could be thought of as what are called the facts of a case, meaning the evidence that is ruled as admissible in a trial. The ground level, on their analysis, consists of a set of arguments. At the argument level, statements can be made about object level statements. In their view, an argument consists of a conclusion and some supporting statements, called premises, linked to that conclusion by some notion of logical consequence. At this level, there is a mechanism for modeling the notion of one argument attacking another one, defined using Bench-Capon's system of value-based argumentation (Bench-Capon, 2003). When this model is applied, an argument attack based on values that the audience holds can be evaluated as weaker or stronger, partly based on these values. At the meta-argument level, an argument analyst can refer to the process by which an argument is established, and discuss and evaluate other properties of it as an argument. As an example, they consider an argument that takes place between advocates in a trial. At the meta-argument level, arguments can be made by the judge about these arguments that took place at the argument level.

Prakken (2001) constructed a formal system to show how shifts in a burden of proof work in legal reasoning of the most common sort. He argued that such questions cannot be answered purely within nonmonotonic logics (Prakken, 2001, 253). He classified such problems as "irreducibly procedural" (253) aspects of defeasible reasoning that can only be modeled by turning to meta-argument level considerations. He illustrated this thesis with the following common type of example (259). Suppose that plaintiff supports her claim that a contract exists by arguing that there was an offer and acceptance by the defendant. She supports this claim by bringing forward two witnesses who testify to her offer and defendant's acceptance. The burden is now on the defendant to question or refute this evidence. Defendant attacks her argument by presenting evidence that the witnesses are unreliable. The burden now shifts to his side, and he risks refutation of the argument based on witness testimony. How should the issue of which side bears the burden of proof be decided? Should defendant have to prove his claim that the witnesses are unreliable? Or should plaintiff have to prove the opposite proposition that the witnesses are reliable? Or to make the case even more problematic, suppose defendant, instead of claiming that the witnesses are unreliable, claims that there is an exception because she was insane at the alleged time of contract acceptance. On which side should the burden to prove or refute this claim lie?

This type of common example of legal reasoning, illustrated by other examples we have examined, including the self-defense example, shows how a burden of proof can shift back and forth in a trial. In Chapter 2, we showed how to model such cases using the Prakken and Sartor system and the Carneades Argumentation System. However, there are cases where ruling on which side should have the burden of proof in cases of such shifts may even require a decision by a judge using argumentation about burden of proof at a metalevel. The general problem posed is to provide a formal framework for allocating burden of proof to a meta-argument level. This is a problem for AI, for law and for applying methods of argumentation and AI to understanding the logical basis of how legal reasoning works. It is also surely a central problem for argumentation theory as a whole, one best approached through the study of metadialogues.

3. Theoretical Problems of Metadialogues

The formalism of Wooldridge, McBurney and Parsons (2005) is a meta-logic, that is, a first-order logic whose domain includes sentences of an object language. Such languages can be self-referential, that is, they can refer to themselves. However, this property is barred because of the problems and paradoxes it can introduce. Instead, Wooldridge, McBurney and Parsons (2005) construct a first-order hierarchical metalanguage such that no sentences from a higher level can be contained in the domain of a lower level. One of the problems with applying such a formal logical structure to any realistic cases of argumentation concerns the interface between levels when there may be a shift to a higher level, and then back to a lower level. It is just this kind of shift that typically occurs in cases of disputes about burden of proof. As the argumentation at one level founders because it has become deadlocked by a dispute about which side should have the burden of proof, there is a need to shift to a higher level in order to rule on this issue. This ruling is made by examining the arguments that produced the deadlock in light of standards and rules concerning burden of proof that may have been established at some prior stage of the dialogue. When is such a shift appropriate, and how should the issue be resolved at the higher level? These are the questions posed if burden of proof is to be studied by applying a metalogic approach. But there are other questions as well. Once the shift has taken place, and some argumentation at the second level has come to some resolution of the issue that blocked progress at the first level, how should the transition back to the lower level take place? And how should any ruling that might have been made at the second level be applied to the argumentation that took place at the first level? Clearly rules are needed to govern such shifts back and forth between higher and lower levels if the shift to the higher level is to be considered reasonable by both parties, and if a solution that has been worked out at the higher level is to be accepted by both of them and fitted back into the lower level so that the argumentation at that level can move forward.

Krabbe (2003, 83) formulated the central problems that arise in connection with metadialogues. One is the demarcation problem of deciding which critical moves belong to the ground level and which ones belong to the meta-dialogue level. Many moves at the ground level may ask for conversational repairs of some sort, but for that reason alone may not need to be classified

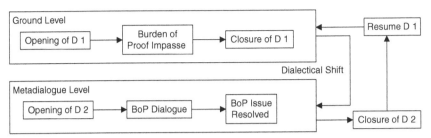

FIGURE 6.1. Shift from a Ground Level Dialogue to a Metadialogue

at the metalevel. However, some serious problems can occur at ground level that might actually block the dialogue from achieving any further progress at that level. A burden of proof impasse is such an example. Suppose one side says, "You prove it!" and the other side replies, "You disprove it."[1] Trying to resolve such a dispute within a dialogue, like a persuasion dialogue, for example, might prove futile. Help is needed, possibly in the form of intervention by a third party, or by some rule or mechanism that is not part of the persuasion dialogue itself, but is needed to resolve the burden of proof dispute. The diagram in Figure 6.1 shows how the need for such an intervention involves a shift from the original dialogue D1 at a ground level to a metadialogue that needs to take place at a different level D2.

The demarcation problem in such a case of a shift to a metadialogue involves examining the textual details of a given case of argumentation, and applying dialogue identifiers to determine the two types of dialogue involved, their stages, and judge where each of the points in Figure 6.1 can be found in the text of the case. But other normative problems also arise. How could we judge whether the shift was an embedding, that is, whether D2 is embedded in D1 so that D2 is really a help in contributing to the fulfillment of the goal of D1 by moving D1 forward in a positive way? Generally, the problem is one of formulating the mechanisms and rules that license such an embedding.

Krabbe (2003, 83) also stated two other central problems. The problem of infinite regress is that a discussion about ground level rules may open up a discussion about rules governing the ground level rules, which might lead to a discussion about the application of the second level rules. The problem is how to block such a regress. The conversation could cycle endlessly back and forth between D1 and D2, and also involve discussions about normative issues of whether the shift is legitimate or improper. The equity problem is that of resolving a metadialogue dispute while blocking unwarranted charges or procedural quarreling.

---

[1] Many examples of this sort can be found in studies of the *argumentum ad ignorantiam* fallacy (Hamblin, 1970; Krabbe, 1995; Walton, 1996).

To analyze the structure of the argumentation in cases where there is a problem of burden of proof, Prakken, Reed and Walton (2005) formalized a protocol for the embedding of burden of proof dialogues into conventional persuasion dialogues. The protocol uses a formal dialectical framework that regulates how the intervening burden of proof dialogue should be embedded into the prior dialogues. To ensure an orderly transition, each move in a dialogue is assigned a level, and there are rules that determine how a given dialogue can move to a higher level in which a burden of proof discussion can take place, and can then move back to the lower level once this metadialogue sequence is terminated. Notice that such a move to a higher level can involve a shift from one type of dialogue to a different type. For example, the initial dialogue might be a persuasion dialogue, but then during the burden of proof interval it may shift to a negotiation dialogue. Once the issue of which side should have the burden proof has been resolved, the argumentation would then shift back to the original persuasion dialogue, and continue where it left off.

4. Analyzing the Los Gatos Example Using Carneades

Now we can return to the example describing the Dutch Supreme Court decision (HR 19 September 1980, NJ 1981, 131) concerning the labor dispute between the band Los Gatos and the Holland America Line. To take a careful look at the argumentation in this case we quote the description of it given in Prakken, Reed and Walton (2005, 116).

A music band named Los Gatos was hired to work on a cruise ship of the Holland-America Line (HAL). At some point the manager told the band to perform for the crew while the ship was waiting for repair in a harbor without passengers. The band refused to play, after which they were immediately dismissed. According to Dutch law, such a dismissal is valid if and only if there was a "pressing ground" for dismissal. One such pressing ground is when the employee persistently refuses to obey reasonable orders of the employer (Section 1639 p.,10 Dutch Civil Code). Los Gatos sued HAL on the ground that this pressing ground would not apply in their case. Their main argument was that the HAL managers had not wanted to listen to the reason why the band had refused to play. This fact was not disputed. What was disputed is how much had to be proven by Los Gatos to claim, that in their case, their refusal to obey the orders of HAL was not a pressing ground for dismissal.

As noted in the presentation of this example in Chapter 2, the issue is one of burden of proof. The nature of the disagreement about burden of proof is clarified in the further remarks on the case given in (Prakken, Reed and Walton, 2005, 116–117).

In particular, the dispute revolved around the issue whether Los Gatos had to prove that they had a good reason to refuse to play or that HAL had to prove the opposite. The Supreme Court decided that HAL had the burden of proof because its managers had made it impossible for Los Gatos to explain their reasons for not wanting to play. Arguably, the underlying dispute was whether the interpretation rule is (3) if employees were not heard then refusal of work is not a pressing ground for dismissal (with an exception for when the employees had no good reason for their refusal) or (3') if employees were not heard and they had a good reason to refuse work, then refusal of work is not a pressing ground for dismissal.

The dispute concerning burden of proof in this case is a subtle one because one has to think very carefully about the difference between the two interpretation rules 3 and 3' to appreciate how this decision affects the outcome of the case. To analyze the argumentation in the case by applying the two burden of proof technologies studied in Chapter 4 to its analysis, we begin by applying the Carneades Argumentation System to it first, and then seeing how the abstract argumentation technology used in the Prakken and Sartor model of burden of proof can be applied to it. This will prove to be a good way to proceed, because the Carneades argument visualization tool, when applied to the argument in this case, provides a clear and easily understandable representation of the basic structure of the argument. Using this method, we can show very clearly how the burden of proof shifts during the sequence of argumentation in the case. Then we can go on to apply the Prakken and Sartor method to it, and that will bring out additional insights on some of the central features of the argumentation in the case and raise some interesting questions concerning the relationships between these two models and general use of dialogue models and methods in argumentation studies.

First then, let us look through the argumentation in the case and see how the basic structure of it can be represented using a Carneades argument diagram. The issue is whether the dismissal of Los Gatos was valid or not. Let us begin in Figure 6.2 by representing the basic argument of HAL to support the conclusion that the dismissal is valid.

We were told in the description of the case that according to Dutch law such a dismissal is valid if and only if there was a pressing ground for it. This rule appears as the first premise in the top argument at the left for the conclusion that the dismissal is valid. The other premise required to support this conclusion is the claim that there was a pressing ground for the dismissal. Holland America gave grounds for supporting this premise by arguing that their employee refused to obey a reasonable order. It was stated in the description of the case that according to the Dutch Civil Code, such a pressing ground obtains when the employee persistently refuses to obey a reasonable order of the employer. Hence, we have put in an additional premise stating the legal rule that if an employee refuses to obey a reasonable order that constitutes a pressing ground. So far, Figure 6.2 represents the

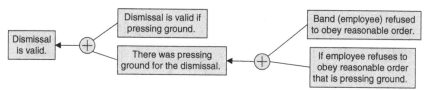

FIGURE 6.2. Basic Argumentation Structure of the Los Gatos Case

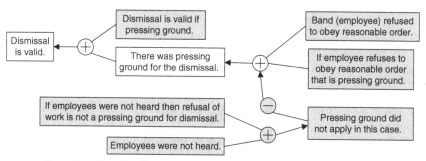

FIGURE 6.3. Step 1 in the Evaluation of the Argumentation of the Los Gatos Case

structure of the pro argument on the side of Holland America for its conclusion that the dismissal is valid.

Now, if we look along the bottom of Figure 6.3, the contra argument put forward by Los Gatos is represented. It is represented in the Carneades Argumentation System as an undercutter that defeats the prior argument of Holland America by claiming that there is an exception to the rule in this case. It is alleged that the "pressing ground" rule does not apply in this case, and offers premises in a pro argument to support this claim. These two premises are shown at the bottom left of Figure 6.3.

In Figure 6.3, the two premises in darkened boxes at the top right of the diagram are shown as accepted. The bottom premise is a rule from the Dutch Civil Code stating that if the employee persistently refuses to obey a reasonable order their refusal constitutes a pressing ground for dismissal. The top premise can also be accepted. We are told in the description that the band refused to play. Nowhere in the case description does it say that the order to play was reasonable, and so we could perhaps have broken this premise down into two separate premises, one of which states that the order was reasonable, while the other states that the band refused to obey it. However, because it does not say anywhere in the case description that the reasonableness of the order was an issue at this point, we have not done this. But now, even though these two premises are accepted, the conclusion that there was pressing ground for the dismissal is not acceptable. Now because it is not the case that both premises of that argument are accepted, the ultimate conclusion that the dismissal is valid is no longer acceptable. At this point then,

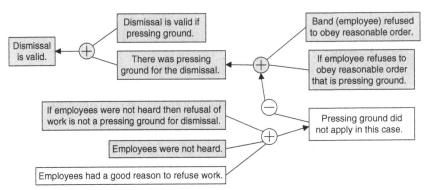

FIGURE 6.4. Step 3 in the Evaluation of the Argumentation of the Los Gatos Case

the burden of production is on Los Gatos, until they put forward further argumentation.

In Figure 6.3 we see what happened in the evidential situation when the counterargument of Los Gatos was put forward. It tells us the main argument of Los Gatos in the description of the case was that the Holland American managers had not wanted to listen to the reason why the band refused to play, and that this fact was not disputed. What was disputed was how much Los Gatos had to prove to claim that their refusal to obey orders was not a pressing ground for dismissal. The question was which of the two interpretation rules 3 or 3' should be applicable. Rule 3 is represented as being the applicable rule in Figure 6.3. This rule is shown in a darkened box in Figure 6.3.

Now if we look at Figure 6.3, we can see how the undercutter defeats the Holland America argument at the top. Both premises of the undercutter argument are accepted, and therefore the conclusion that the pressing ground does not apply in this case is also shown in a darkened box. Thus, even though both premises of the argument supporting the conclusion that there was a pressing ground for the dismissal are accepted, and shown in darkened boxes, they still failed to provide adequate support for the conclusion because the argument from the premises to the conclusion are undercut by the Los Gatos contra argument shown at the bottom. Therefore, the premise that there was a pressing ground for the dismissal is now shown in a white box, revealing that the ultimate conclusion (that the dismissal is valid) will also automatically be shown in a white box by Carneades. This way of interpreting the rule, as expressed by 3, means that the argument of Holland America is now defeated.

Now finally, let us turn to Figure 6.4. Figure 6.4 implies the interpretation rule 3', which says that if employees were not heard and they had a good reason to refuse to work, refusal of work is not a pressing ground for dismissal. We can represent this way of interpreting the argument by adding an additional premise to the Los Gatos argument shown at the bottom of Figure 6.4.

This way of visualizing the argument puts the burden of proof on Los Gatos to prove that they had good reason to refuse work. Given the description of the case, there is no evidence supporting this claim. The outcome is that this premise is shown in a white box at the bottom of Figure 6.4. This means that one of the premises is missing in the Los Gatos argument to prove that a pressing ground did not apply in this case. This in turn means that the undercutting argument fails, and that once again the argument of Holland America shown at the top with all darkened boxes proves its conclusion that the dismissal was valid.

To see the differences between the two rulings on burden of proof in the case, you have to examine the difference between Figure 6.3 and Figure 6.4. Figure 6.3 puts the burden of proof on Los Gatos to prove that they had good reason to refuse work. This ruling on burden of proof means, given the evidence shown in the case, Holland America's conclusion that the dismissal is valid is proved. The other way of ruling on the burden of proof is shown in Figure 6.4, displaying the line of argumentation that shows that the claim that the dismissal is valid is not proved.

What is shown in this case is that the judge had to make a decision on how best to represent the proper sequence of argumentation in the Los Gatos case. Should the argumentation in the case be properly represented by the diagram in Figure 6.3, or by the diagram in Figure 6.4? This decision depends on whether interpretation rule 3 or interpretation rule 3' is chosen as the major premise of the undercutting argument put forward by Los Gatos to the effect that "pressing ground" does not apply in this case. Looking at Figures 6.3, you can see the two different outcomes of the two choices in this decision. In Figure 6.3, the ultimate conclusion that the dismissal is valid appears in a white box, showing that it is not proved. In Figure 6.4, the ultimate conclusion is visually represented in a darkened box, showing that it is proved.

Because the judge had to make a decision on which interpretation rule should be applied to this case, thereby setting in place the burden of proof that would determine the outcome of the case, the determination of burden of proof can be seen as external to the argument itself. This determination requires an ascent to a higher level of argumentation where the judge had to arrive at a decision on which way to interpret the argumentation. Thus, the case can be classified as one where there needed to be a higher-level dialogue to arrive at a determination of the burden of proof to be applied in the first level dialogue representing the argumentation on both sides in the trial.

5. Analyzing the Los Gatos Example Using
Abstract Argumentation

Prakken, Reed and Walton (2005) have reconstructed the burden of proof debate from this case using a formalized protocol for burden of proof dialogues. The protocol is formulated as a formal dialogue game with commitment rules that define the preconditions and postconditions of the speech acts used by both parties in the dialogue. The type of dialogue is that of a persuasion dialogue, and hence, a basic persuasion protocol is used in which the burden of proof is hard-wired into the system. Relevance is defined in terms of the dialectical status of a move, which is in turn recursively defined by the nature of the replies to the move. A move is said to be *in* if it is surrendered or if all of its attacking replies are *out*, meaning that a move without replies is *in*. A move is *out* if it has an attacking reply that is *in*. Finally, a move is *definitely in* or *definitely out* if it is in or out and its status cannot change anymore. Another requirement is that a move is *surrendered* if it is an argument move and it has a reply saying that the argument is conceded or if it has any surrendering reply. These rules enable us as critics to determine by looking at the argumentation stage in any given case how a move is in or out, and how each subsequent move becomes in or out based on the rules and on the previous moves. Using this formal dialogue protocol Prakken, Reed and Walton (2005) presented the following reconstruction of the burden of proof debate from the Los Gatos case. *P* is the proponent in the dialogue (Los Gatos) and *O* is the opponent (Holland America). The moves are numbered, and the speech acts at each move are labeled in a way that indicates what type of move is made. The target of a move, the move to which it was addressed, is shown in the second column. The metadialogue of the Los Gatos case is from (Prakken, Reed and Walton, 122) presented in Table 6.1 as a profile of dialogue (Krabbe, 1999).

At move O10 the burden of proof dialogue is terminated, so that the transition back to the original persuasion dialogue can be made. The burden of proof dispute having been resolved in this interval, the argumentation in the original case can be carried on. The burden of proof dispute arises when the opponent asks at O6 why there is a good reason for refusal, and the respondent replies at O7 by asking why there is not a good reason for refusal. But then at O8, the opponent invokes the notion of burden of proof, using the normal default rule to argue that the plaintiff must offer a reason to prove his main claim. This means that O6 is now in, and that a good reason for refusal must be provided. The proponent fulfills this BoP request by providing such a reason. He argues that the employer made expressing reasons for refusal impossible.

This analysis of the shifting of the burden of proof in the Los Gatos case is especially interesting from the point of view of the theory of burden of proof in argumentation studies because it has a dialogue format showing the burden of proof shifting from one side to the other. What is also very interesting, however, is that the argumentation in the example can also be displayed using the Dung-style abstract argumentation method explained and applied to a different case in Chapter 4. The argumentation in the Los Gatos example is shown in Figure 6.5 using this method.

TABLE 6.1. *Embedded Metadialogue in the Los Gatos Case*

Move	Target	Speech Act and Content	Effect
P1	nothing	*claim* dismissal-void	P1 is in
O2	P1	*why* dismissal-void?	P1 is out, O2 is in
P3	O2	dismissal-void *since* no pressing-ground	O2 is out, P1 is in
O4	P3	*why* no pressing-ground?	P3 and P1 are out
P5	O4	no pressing-ground since not-heard	O4 is out, P1 is in
O6	P5	*why* good-reason-for-refusal?	P5 and P1 are out
P7	O6	*why* not-good-reason-for-refusal?	P1 is in, O6 is out
O8	P7	*BoP* (good-reason-for refusal, P), *since* plaintiff must prove his main claim	P7 and P1 are out, O6 is in
P9	O8	*BoP* (not-good-reason-for refusal, O) since employer made expressing reasons for refusal impossible	O8 is out, P1 is in
O10	P9	*concede* BoP (not-good-reason-for refusal, O)	P9 is definitely in, O8 is definitely out, P1 is in

At move one, the proponent's ultimate claim that the dismissal is void is in. At move two, where the opponent attacks the proponent's initial claim by asking the question why the dismissal is void, the proponent's initial claim is now shown as out. This means that the burden of proof is on the proponent to back up his claim that the dismissal is void by means of evidence, and until he does so, his claim is to be regarded as unproven and is therefore not accepted. At move three, the proponent responds to the question by giving an argument to support his claim that the dismissal is void. He argues that the dismissal is void because there is no pressing ground. And so forth, if we track down the sequence of moves in Figure 6.5 from move one to move ten, each move defeats each prior move that was its target. Ultimately this process of argument and defeat reaches the final move, move ten, where the proponent's argument is in.

What we have seen then is that either system, the Carneades Argumentation System or the abstract argumentation framework employed by Prakken and Sartor, can be used to model burden of proof in examples like the Los

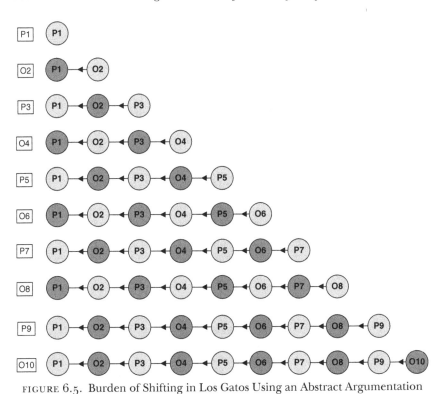

FIGURE 6.5. Burden of Shifting in Los Gatos Using an Abstract Argumentation
Sequence

Gatos case to explain how the burden of proof shifts back and forth over an extended sequence of argumentation.

This case also shows how a burden of proof metadialogue can be embedded into an original persuasion dialogue, and used to solve a hard problem about burden of proof so that the original dialogue can continue in a constructive manner toward its goal. In this case, the main issue in the trial was that of burden of proof, and hence, the metadialogue reconstructed is the tool that shows how the dispute in the trial as a whole should be resolved. This case is an especially interesting one because it shows not only how a dispute about burden of proof can be the main issue in a trial, but also how either computational framework for resolving the issue requires the embedding of a metadialogue into the original dialogue.

6. Arguments from Ignorance Revisited

In legal argumentation, argument from ignorance is closely associated with what is often called the presumption of innocence in a criminal trial. This terminology is misleading, as pointed out in Chapter 2. It should be said that the burden of persuasion is on the prosecution side. The prosecution has the burden of proof and must bring forward enough evidence to satisfy the proof standard of beyond reasonable doubt. The defendant need only bring forward enough evidence to prevent the prosecution from meeting its burden of proof, by casting enough doubt on the prosecution's attempt to prove its claim. This asymmetry involves an argument from ignorance. If the defense can show that there is a lack of evidence to support the prosecution's claim (ultimate thesis to be proved in the trial), then the defense has shown that this claim does not hold up and must be rejected. This form of argumentation meets the requirements for the argumentation scheme of the argument from ignorance. Thus, argument from ignorance is fundamental to the argumentation structure of the trial in the adversary system.

Argument from ignorance is also common in more special forms in legal argumentation. For example, as shown by Park, Leonard and Goldberg (1998, 103), there is a presumption that some writing has been accurately dated: "unless the presumption is rebutted, the writing in question will be deemed accurately dated." Another example (153) concerns character evidence. Suppose a first person was in a position to hear derogatory statements about a second person if any were made. And suppose the first person testifies that he heard no such comments. This testimony counts as evidence of the second person's good character. The form of argument in such a case is that of argument from ignorance. If no evidence of bad character was found or reported by the witness, this lack of such a finding may be taken as evidence of good character.

But are such arguments from ignorance fallacious, or even suspicious and deceptive, as Gaskins maintains? To address this question we have to examine the logical form (the argumentation scheme) of this type of argument.

The simplest formulation of the scheme for the *argumentum ad ignorantiam* of the logic textbooks is this: statement A is not known to be true (false), therefore A is false (true). The following example can be used as an illustration of this type of argument. In this particular case,[2] a guest on the "Antiques Roadshow" presented a Colt model 1849 pocket revolver given to his great-great grandfather in about 1872. The gun allegedly belonged to a bodyguard for President Lincoln who supposedly got a shot off at John Wilkes Booth during the assassination. As supportive evidence he presented a letter from his great-great grandfather stating that the pistol had been used by a bodyguard of Lincoln's to shoot Booth in the leg, breaking the leg. Evidence of the date of manufacture of the gun showed the timeline was possible. The dialogue quoted from the transcript ran as follows.

---

[2] http://www.pbs.org/wgbh/roadshow/archive/200705A13.html#.

Appraiser: At that point. In this condition, at auction, I would estimate this pistol is worth about $2,000 to $3,000.

Guest: Oh, wow.

Appraiser: If there was any way that we could truly document the history behind it and support it more, I would guess it would be about $15,000 to $20,000.

Guest: Oh, wow.

Appraiser: And, um, part of this is that in many cases, family histories become cloudy.

Guest: Exactly.

Appraiser: Even if he was there, and did take a shot, it would have been big news. I mean ...

Guest: You would think so, exactly.

Appraiser: This is one of the most investigated crimes in its day, so if he had any sort of even peripheral association, it would have been documented.

Guest: Well, thank you.

Appraiser: You're welcome.

In this case, it makes a great difference to the value of the Colt revolver whether it was a gun used by one of Lincoln's bodyguards to cause an injury to the leg of John Wilkes Booth after he assassinated Lincoln. The case supporting the claim that this revolver was so used is weak, however, because as the appraiser put it, "family histories become cloudy." The generally accepted opinion is that Booth injured his leg by jumping from Lincoln's box at the theater onto the stage before he escaped through an exit. The general burden of proof, or what might be called the burden of persuasion in this case, even though it is a case from everyday conversational argumentation and not specifically a legal case, should therefore be set at a fairly high standard of proof. In order to prove the claim made by the guest on the "Antiques Roadshow" that this gun was used to injure Booth, some fairly strong and convincing evidence would have to be presented. Argument from ignorance can be used to support this way of framing the argument.

Argument from ignorance, or argument from lack of evidence, as it might better be called, has the following argumentation scheme (Walton, 1996, 254).

Major Premise: If A were true, A would be known to be true.
Minor Premise: A is not known to be true.
Conclusion: A is false.

The major premise assumes that there has been a search through the knowledge base that would contain A that has supposedly been deep enough so that if A were there, it would be found. The critical questions include considerations of (1) how deep the search has been, and (2) how deep the search needs to be to prove the conclusion that A is false to the required standard of proof in the investigation.

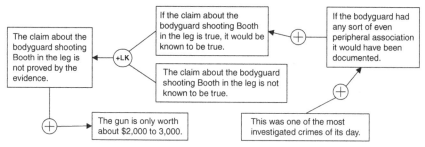

FIGURE 6.6. Carneades Argument Diagram for the Colt Revolver Example

We can apply the scheme for argument from lack of evidence to this case as follows. The major premise put forward by the appraiser is that if the bodyguard in question was there at the scene in Ford's Theater and did take a shot, it would have been big news because this was one of the most investigated crimes of its day. In other words, the appraiser was supporting the major premise of the scheme, namely the proposition that if the claim about the bodyguard shooting Booth in the leg were true, it would be known to be true. The way he is using the argument from ignorance is displayed by the argument diagram in Figure 6.6. LK represents the scheme for argument from lack of knowledge, or argument from ignorance as it is often called.

Now it is true that the argument used in this case is conjectural. It could be true that the injury to Booth's leg was caused by the bodyguard's firing this Colt revolver. The argument from ignorance used by the appraiser does not rule out that possibility altogether. However, it does shift the burden of proof against that hypothesis, on the balance of the evidence and lack of evidence. This way of evaluating the argumentation in the case suggests that the argument from ignorance is reasonable when used within its limits. But if the reader is not convinced by an everyday conversational argument used on the "Antiques Roadshow," another example might be cited of the same form of argument used in an academic discipline, namely history

As noted earlier, this form of argument is often called the lack of evidence argument in the social sciences or an *ex silentio* argument in history, where it is presumed not to be fallacious. In both fields it is commonly regarded as a reasonable but inconclusive form of argument. To cite an example (Walton, 1996), there is no evidence that Roman soldiers received posthumous decorations, or medals for distinguished service, as we would call them. We only have evidence of living soldiers receiving such awards. The argument is shown in Figure 6.7.

From this lack of evidence, it has been considered reasonable by historians to put forward the hypothesis that Roman soldiers did not receive posthumous decorations. Of course, such a conjecture is not based on positive evidence, but only on a failure to find evidence that would refute it. But still, the

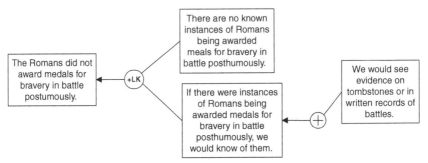

FIGURE 6.7. Carneades Argument Diagram for the Roman Medals Example

argument seems reasonable, even though it is not conclusive, but only holds as acceptable based on a balance of positive versus negative evidence.

As shown by Macagno and Walton (2011) there is a growing body of literature arguing that use of negative evidence is both useful and necessary in science. Smets (1991) has shown how establishing well-founded theories in science rests on arguments from lack of evidence, combined with positive evidence in an investigation. Witte, Kerwin and Witte (1991) have provided many examples of reasoning from negative evidence in medical education and research. Dohnal (1992) maintains that arguments from negative evidence are common and useful in reliability reasoning used for testing microelectronic circuits. In Walton (1996) it is argued that negative evidence can be used to guide scientific inquiry. The conclusion suggested by this data is that argument from ignorance can be a reasonable kind of argumentation when used in an appropriate setting of research and inquiry under the right conditions.

As shown in (Walton, 1996), argument from ignorance needs to be seen as an inherently dialectical form of argumentation. The context most often helpful to grasping the structure of this form of argumentation is that of an ongoing investigation in which facts are being collected and inserted into a knowledge base. In such a context, the argument from ignorance can be represented using the following argumentation scheme for epistemic argument from ignorance given in (Walton, Reed and Macagno, 2008, 328). In this knowledge-based scheme, D is a domain of knowledge and K is a knowledge base in a given domain, or field of knowledge.

All the true propositions in D are contained in K.

A is in D.
A is not in K.
For all A in D, A is either true or false.
Therefore, A is false.

This form of argument can be deductively valid in a domain D where K is closed, meaning that it contains all the statements that can ever be known in that domain. But in a vast majority of cases, argument from ignorance is a defeasible inference that may default as an investigation proceeds and new knowledge comes in. Thus, one of the most important critical questions in evaluating any given instance of an argument from ignorance is whether the knowledge base is open or closed.

In many instances, the shifting of the burden of proof can be detected in a dialogue format in which the proponent puts forward a request that the respondent should prove something and the respondent, instead of answering with the requested proof, demands that the proponent prove the opposite. Such a shifting of the burden of proof dialogue, as discussed by Krabbe (1995, 256), takes the following form.

Proponent: Why A?
Respondent: Why not-A?

The classic example was the case outlined in chapter 1, a political debate in which a member of the opposition party demanded that the government minister should prove with absolute certainty that Canadian uranium was not being used for military purposes. In this example, the respondent, the government minister, replied that the opposition should give evidence to support the allegation that Canadian uranium was being used for military purposes. The dialogue exchange characteristic of this type of argumentation will be easily recognizable to many readers as an instance of the argument from ignorance, the so-called *ad ignorantiam* fallacy. It is interesting to observe that this fallacy can be identified in a dialogue format where there is a reversal of burden of proof from one move to the next.

An example that can be used to illustrate how the nonfallacious *ad ignorantiam* works as an argument is the foreign spy argument: Mr. X has never been found guilty of breaches of security, or of any connection with agents of the foreign country he is supposedly spying for, even though the Security Service has checked his record; therefore, Mr. X is not a foreign spy. This argument from ignorance is defeasible because it is not possible to be absolutely certain that Mr. X is not a foreign spy. Mr. X could have avoided detection through many security searches. He could be a "mole," a deeply hidden spy who has never been discovered. The question then is when can a search be closed on the basis that enough evidence has been collected.

There is a very common principle often appealed to in knowledge-based systems in AI called the closed world assumption (Clark, 1978). Essentially, the closed world assumption means that all the information that there is to know or find is listed in the collection of information one already has, but there are different ways of representing information. According to Reiter (1980, 69), the closed world assumption is met if all the positive information in a database is listed, and therefore negative information is represented by default. Reiter (1987, 150) offers the example of a database for an airline flight schedule to show why negative information is useful. It would be too much information to include in such a database all flights and the city pairs they do not connect. This amount of information would be overwhelming. Instead, the closed world assumption is invoked. If a positive flight connection between a pair of cities is not asserted on the screen representing the database, the conclusion is drawn that there is no flight connecting these two cities. If the system searches for a flight of the designated type and does not find one in the data base, it will reply "no." Reiter (1980, 69) described the form of argument used in this sequence of knowledge-based reasoning as: "Failure to find a proof has sanctioned an inference." As noted earlier, this kind of inference by default from lack of knowledge has traditionally been called the *argumentum ad ignorantiam* in logic. The argument from negative evidence may be merely a defeasible inference that leads to a provisional commitment to a course of action, but should be seen as open to new evidence that might come into an investigation and needs to be added to the knowledge base.

7. The Nonfallaciousness of Argument from Ignorance

In Chapter 1 (Section 2) it was explained how Gaskins (1992) linked argument from ignorance to the way burden of proof, he claimed, is used in law as a fake and shadowy decision rule for drawing inferences from lack of knowledge. He describes the use of this form of argumentation as a device for giving stage directions in legal argumentation where one party has an obligation to speak first, and then the second party can step forward with evidence. By associating it with the traditional fallacy of argument from ignorance, a very negative sounding form of argument, Gaskins portrays burden of proof as a shadowy tool of legal argumentation that lawyers and judges can wield in the background in order to manipulate litigation. His claim that the notion of burden of proof is closely related to the form of argumentation traditionally known as argument from ignorance in logic is quite right. Argument from ignorance is often taken in logic to be equivalent to shifting the burden of proof in a dialogue. However, the conclusion that Gaskins draws from this equivalence is erroneous, depending as it does, on the additional assumption that argument from ignorance is a fallacy, despite the acceptance of this assumption in traditional logic (Walton, 1996, Chapter 2). The examples in Section 6 show that argument from ignorance is, in many instances, a reasonable but defeasible form of argument that can be used in a fallacious manner, but can be used appropriately in argumentation to shift a burden of proof in many other instances.

The simplest formulation of the scheme for the *argumentum ad ignorantiam* of the logic textbooks is this: statement A is not known to be true (false), therefore A is false (true). This seems like it could be a fallacious move because negative evidence, in the form of a failure to show that a particular proposition is true, does not conclusively demonstrate that this proposition is false. Suppose we take deductive logic to be the model, and represent this kind of argument as one that needs to meet the standard of deductive validity. On this view, argument from ignorance seems like a logical leap that makes it always fallacious. However, once we start to study examples of this form of argumentation in scientific fields like history and law, and in everyday conversational argumentation, as defeasible arguments that are extremely common, we begin to find many examples of argument from ignorance that are reasonable to shift a burden of proof, even if they are not conclusive forms of argument. In the Roman medals example, through many years of historical investigations no evidence was ever found that Roman soldiers received posthumous decorations, or medals for bravery. Based on this lack of evidence, the hypothesis was tentatively put forward by historians of the ancient world that Roman soldiers did not receive posthumous decorations. Provided the historian takes it as a defeasible argument that is open to refutation in the future should new contrary evidence come to be discovered, it seems like a reasonable argument. It is a negative form of argument that is inherently defeasible because the hypothesis formed in this way is not based on positive evidence, but only on a failure to find any evidence that refutes it. Such arguments from ignorance are common in many fields, not least in law, as will be shown later.

We now know that there is a defeasible argumentation scheme for argument from ignorance, and that if a particular argument fits the scheme and meets all its requirements it can be a reasonable argument used to shift the burden of proof from one side to the other in a dialogue.

The scheme for argument from ignorance can be seen to be an instance of a special type of defeasible *modus tollens* argument, assuming that the rule of double negation holds, stating that A is false if and only if A is not true. How it works as a reasonable form of argument can be shown by looking at the argument diagram in Figure 6.7. The conclusion, shown at left, is the negative statement that the Romans did not award medals for bravery in battle posthumously. The two premises on the right of the node containing the name of this scheme "argument from ignorance" are used in a pro argument supporting the conclusion.

Note that the diagram in Figure 6.7 displays some evidence backing up the conditional statement that is one premise of the argument from ignorance. This evidence is supplied by the statement that we would see evidence on tombstones or in written records of battles. Because this particular argument fits the scheme for argument from ignorance, and because the conditional premise is supported by evidence of visual records and other data known to historians, it appears to be a reasonable argument, even though it does not prove its conclusion to the standard of beyond all doubt. It is a defeasible argument, but one that gives us a perfectly good reason to tentatively accept the conclusion as an item of historical knowledge.

The reason for the tendency to think of arguments from ignorance as fallacious may be that Pascalian probabilistic reasoning assigns probabilities to all the propositions being considered when making an evidential judgment, and therefore does not take ignorance into account (Stein, 2005, 45). In the Baconian framework, in contrast, a proposition supported by a large amount of evidence is evaluated differently from a proposition resting on only a small amount of evidence. Baconian probability takes two factors into account in judging the evidential weight of a hypothesis (Anderson, Schum and Twining, 2005 259). One is how much evidence has been collected so far that is favorable to the hypothesis. The other is how many questions regarding the acceptability of the hypothesis remain unanswered by the evidence collected so far. In other words, Baconian probability takes into account not only what we know, but also what we don't know, when judging the acceptability of a hypothesis. Thus, there is a framework of probability that not only assumes that the argument from ignorance is in many instances a reasonable form of argument, but that requires consideration of the closed world assumption as an important factor for evaluating probable reasoning. Within the Baconian framework, the argumentation scheme for argument from ignorance is a vital tool for evaluating probabilistic reasoning used to judge the evidential support of the hypothesis.

Even though a knowledge base is incomplete, and the search for new knowledge may still be underway, this scheme can still enable a conclusion to be tentatively drawn by defeasible reasoning. In such an instance, the argumentation scheme becomes a defeasible form of argument, holding only tentatively, subject to the asking of critical questions during a search for more knowledge that may continue. The first premise described earlier is associated with the assumption that there has been a search through the knowledge base that would contain A that has been deep enough so that if A were there, it would be found. One critical question is how deep the search has been. A second is the question of how deep the search needs to be to prove the conclusion that A is false to the required standard of proof in the investigation. It is not necessary to go into all the details here, given space limitations, but enough has been said to draw a parallel with the analysis of argument from expert opinion described earlier.

Arguments from ignorance having the form of the argumentation scheme set out in Section 6 are best analyzed as defeasible arguments at some stage of a dialogue or investigation in which evidence is being collected and assessed. The pattern of reversal of burden of proof characteristic of the argument from ignorance is displayed in this characteristic sequence of dialogue in which there is a shifting of the burden of proof identified by (Krabbe, 1995, 256).

Proponent: Why A?

Respondent: Why not A?

Indeed, it was just this problem that was identified in Chapter 1 by the example of the political debate where the opposition party asked the government minister to prove with absolute certainty that Canadian uranium was not being used for military purposes. The minister replied that the opposition should give evidence to support their allegation that Canadian uranium was being used for military purposes. In such a case, we have an illustration of the use of the argument from ignorance as a tactic used by both sides in a tug-of-war to try to evade the burden of collecting evidence that would resolve the dispute. This is certainly a common kind of problem associated with the argument from ignorance, and with burden of proof generally, but it does not follow from this kind of example that argument from ignorance is always used fallaciously, or is somehow an inherently erroneous and deceptive form of argument.

Indeed, arguing from negative evidence is a form of reasoning commonly used in knowledge-based systems in AI where it is associated with the use of what is called the closed world assumption. The closed world assumption is said to be met if all the positive information in a database is listed, and therefore negative information is represented by default. As indicated on page 195, Reiter (1987, 150) used the common example of a database of the kind one would see displayed in a monitor in the airport listing the airline flight schedule for that airport. If a flight connection between a pair of cities is not listed on the screen, the conclusion is drawn that there is no flight connecting these two cities. Reiter (1980, 69) described the form of argument used in this sequence of knowledge-based reasoning as an argument from negative evidence: "Failure to find a proof has sanctioned an inference." Hence, there is nothing inherently wrong with arguing from lack of evidence to a conclusion, provided the argument is seen as based on the two premises shown in the argumentation scheme for argument from ignorance presented earlier.

Obviously, the burden of proof requires specific standards to be fulfilled that are appropriate for the case and the type of investigation in which the evidence is to be assessed. A negative proof needs to be based on a certain amount of data, and such data must provide enough evidence to meet the appropriate standard of proof. Even when the standard has been met, reasoning from lack of evidence is somehow always an incomplete proof that may later have to be retracted. The burden of providing evidence is unevenly allocated on the proponents of positive or negative conclusions, and in many cases such a burden cannot be reversed only to make the proving procedure simpler. Reasoning from lack of evidence is an instrument of defeasible argumentation to provide a *prima facie* case, and shift a burden of production onto one's opponents. If the other party cannot provide positive evidence to rebut, or an explanation for the negative findings other than the proponent's hypothesis, the conclusion should be held as acceptable for the time being. The solution to the problem of how to regulate such exchanges is found in the dialogue systems described in Chapter 4 where there are protocols governing each move made by each party in a dialogue. Each move has well-defined pre- and postconditions. When a why question is asked by one party requesting evidence to support a claim made by the other party, the postcondition defined by the reply protocol requires that such evidence be given or that the second party give up his claim (with exceptions where the claim can be shown to be based on common knowledge). These protocols have to be appropriate for the type of dialogue underway. Under the right conditions, proved failure to find evidence can itself qualify as a kind of evidence used in answer to the why question.

It has been shown in (Macagno and Walton, 2011) why drawing inferences based on negative evidence is a rational method of argumentation that should be taken into account when drawing conclusions based on positive evidence. Negative evidence has often been discounted as being of less worth than positive evidence. But this attitude toward it is an imbalance that has often led to biased and misleading research results in scientific inquiries.

In legal argumentation, as pointed out in Chapter 2, argument from ignorance has traditionally been closely associated with what is called the presumption of innocence in criminal law. This terminology is misleading. As pointed out in Chapter 2, the proper way to speak about this matter is to say that in a criminal case the burden of persuasion is on the prosecution side. Argument from ignorance is also related to presumption in legal argumentation. For example, in Park, Leonard and Goldberg (1998, 103), there is a presumption that some writing has been dated accurately: "unless the presumption is rebutted, the writing in question will be deemed accurately dated." Another example (153) concerns character evidence. Suppose a first person was in a position to hear derogatory statements about a second person if any were made. And suppose the first person testifies that he heard no such comments. This testimony counts as evidence of the first person's good character. The form of argument these cases take does seem to be properly described in a way that makes it fit the argumentation scheme for argument from ignorance. If no evidence of bad character was found or reported by the witness, this lack of such a finding may be taken as evidence of good character.

8. When Should a Persuasion Dialogue Be Closed?

When should a persuasion dialogue be closed? According to rule nine of the ten dialectical rules (van Eemeren and Grootendorst, 1992, 209), a failed defense of the thesis put forward by one party requires that party to immediately retract the thesis, meaning that this party has lost the dialogue and the other party has won. But in practical terms, in any given case, how can we tell when such a failure has occurred? The participants themselves would presumably dispute this in the case of a two-party persuasion dialogue, but if we had a third-party referee, it is the referee who has to decide when the dialogue is closed. But this just pushes the question back a little further. We have to ask what criterion the referee would use to determine any given case when one of the parties has failed and must therefore retract his or her thesis. There are various candidates for such a criterion that need to be discussed.

The obvious criterion is to close the dialogue when one of the participants comes up with a strong enough argument to prove her thesis. But how strong does such an argument need to be? In general, once a critical discussion has reached its closing stage, and the argumentation on both sides has been summed up, the side that has the stronger argument wins. In other words, the side that has marshaled the most evidence to prove its thesis should be judged to be the winner. This requirement is basically the preponderance of evidence standard. Even although the evidence put forward by one side might be only slightly stronger than that put forward by the opposed side, if we have to make a decision on which side is the winner, the side with the stronger body of evidence should win. The problem with this criterion for closing of the dialogue, however, is that because the winning side's argument may be only slightly stronger than that of the losing side, it is easily conceivable that the losing side, at its next move, might change the balance by coming up with a new argument that now gives it the preponderance of the evidence. If this possibility exists, and it does not seem possible to rule it out generally, it would be a pity to close off the dialogue just at that point.

Another possibility would be to set a higher standard of proof at the opening stage, say a standard that requires that the dialogue be closed off if one of the parties presents an argument that is so strong his or her thesis is proved beyond reasonable doubt by this argument. This kind of proof would seem to give a very good reason for closing of the dialogue, because if the one-party's thesis is now proved beyond reasonable doubt, there is no point of continuing, because on this point the other party cannot be raising a reasonable doubt. However, implementing this criterion for closure in a given case in persuasion dialogue is generally not as straightforward as it seems, because we now have the problem of judging in a particular case when the thesis has been proved beyond reasonable doubt. The other problem is that even if a participant does prove her thesis to the beyond reasonable doubt standard, it may still be possible that the other party might come up with some argument that nobody has anticipated that shows that there remains some grounds for doubting that this participant has proved her thesis. After all, sometimes the argument looks so strong that it is overwhelming, and there seems to be no grounds for doubting it could have been overcome, still, by its nature argumentation can take us down unanticipated paths and we might come up with a new argument that might cast an argument it is directed to into a new kind of doubt. Van Eemeren and Grootendorst (2004, 137) have a rule stating that a participant in a critical discussion always has the right to challenge the argumentation put forward by the other side. They write, "We propose to grant the right to challenge a discussant to defend his standpoint unconditionally to any discussant who has called the standpoint into question." They continue, "This means that in principle there is no restriction on challenging any discussant on any standpoint by any discussant." According to their rule 2 for the critical discussion (137), any discussant who has called the standpoint of the other party into question is always entitled to challenge this party to defend his standpoint. The problem is that even though the one party has put forward an extremely strong argument at some point in the dialogue, an argument that proves its ultimate thesis to such a high standard that can be classified as falling under the category of beyond reasonable doubt, it might be unfair to the other party, given that he has the right to challenge any argument put forward by the other side, to rule that he can't challenge this particular argument. Who knows whether he might not have a strong rebuttal that nobody else has thought of, and it might be unfair to him not to have an opportunity to bring it forward.

Yet another problem is that in many kinds of disputes in everyday argumentation, for example in ethical and philosophical disputes, setting a standard of proof so high as to require that a thesis be proved beyond reasonable doubt would never be realistically applicable to the arguments on both sides of the dispute. The reason is that if the dispute is about a philosophical topic, like free will or the existence of God, it is unreasonable to expect that either side will reach such a high standard of proof so that the other side cannot still cast some doubt on my reasonable arguments.

To deal with these kinds of cases we might bring in the maieutic function (Walton and Krabbe, 1995), and adopt a criterion that the dialogue should be closed off when the argumentation on both sides has gone into sufficient depth on the topic being discussed. For example, suppose that each side has probed critically into the argumentation of the other side, and also presented extremely deep arguments supporting its own side, in response to the criticisms addressed to it, so that the audience judges that the maieutic function has been fulfilled. Of course, we have the problem of determining precisely how the audience arrives at such a judgment. But it could be possible that there could be criteria for judging the depth of the discussion. So this is a possibility that remains to be explored.

If all these criteria turn out to be problematic, or difficult to implement in precise ways, there are some other possibilities. One of these is to time the dialogue. This method is often used in forensic debates, where a referee allots a fixed time to each party to put forward its arguments and make its replies. However, if we are looking at abstract models of persuasion dialogue, these are not taking place in real time. However, the moves are numbered in a sequential order, and this could provide a criterion. We could designate at the opening stage a specific number of moves that each party can have, so that once this number of moves has been run through, the dialogue is closed. This could be a practical method of closure that is useful in some cases, but in general it may not give a realistic or useful criterion of how to close the dialogue by designating one party is the winner and the other is the loser. It may be that each party has been allotted a large number of moves, but they are not getting anywhere, meaning that the argumentation on one side is either no stronger than that on the other, or at any rate is not strong enough to lead to a reasonable judgment that one is definitely the winner and the other is closer. So we seem to be back at the point where we started. The dialogue should be closed off when it has been determined that one is the winner and the other is the loser, but we still seem to have no single foolproof way of deciding when that stage has been reached that can be applied to all cases of persuasion dialogue. It may be then that because there are different kinds of persuasion dialogue in law, in forensic debates, and in everyday conversational argumentation, practical criteria for determining when each type of dialogue should be closed needs to vary with the type of persuasion dialogue, and perhaps also with practical constraints, like time or the cost of continuing the dialogue.

9. A Solution to These Problems

In this section it is shown how the techniques and concepts studied in the previous sections can be put together into a general method for resolving conflicts about burden of proof. It has been shown that there are three kinds of cases. In the first type of case, the dispute can't be resolved by the method, other than hypothetically, because not enough data is given about the argumentation and the context of dialogue in the case. In the second type of case, the dispute can be resolved using the general criteria for setting burden of proof that apply to the case, but without having to use the device of an embedded burden of proof metadialogue. In the third type of case, the dispute can only be resolved by using a burden of proof metadialogue. In addition to providing a tool that can be applied to cases, it is shown in this section how the components of the previous sections fit together, providing a general theory of burden of proof shifts that is a solution to the problem of burden of proof in argumentation of any sort. It is argued, in other words, that the solution applies to all argumentation in which allocation of burden of proof plays a role.

The solution to the problem of burden of proof begins with the recognition that disputes about burden of proof can occur at any one of all the three stages of a dialogue. Most importantly, the global burden of proof is set at the opening stage. At this stage, the participants have to decide what type of dialogue they are supposedly engaging in. For example, it might be a persuasion dialogue, a deliberation dialogue or a negotiation dialogue. Let's say it's a persuasion dialogue. At the opening stage, it needs to be decided which party has the burden of proving or doubting which proposition. This is the stage that van Eemeren and Grootendorst (1992) call the confrontation stage. For example, in a persuasion dialogue there can be two types of conflicts of opinions. In the dispute, one party has a designated proposition to prove, while the other party has the burden of proving the opposite proposition. In the type of dialogue called the dissent, one party has the burden of proving a designated proposition, while in order to be successful in the dialogue, all the other party has to do is to cast doubt on the first party's attempts to prove this proposition. At the opening stage, both parties accept procedural and material starting points. For example, supposing it is a legal dispute, one that needs to be resolved by a procedure like a criminal trial. At this stage, each of the parties needs to fulfill its burden of proof by putting forth arguments and other speech acts. In the case of a criminal trial, for example, that standard is one of proving something beyond a reasonable doubt.

At the argumentation stage the local burden of proof can shift back and forth. What is important for regulating burden of proof, and resolving disputes about burden of proof at this stage, is determined by the protocols for different types of speech acts. Centrally important here are the properties of the speech acts relating to presumption shown in Chapter 3. The speech acts of presumption, assertion, assumption and putting forward an argument are the central ones to be considered in many of the most common kinds of cases of disputes about burden of proof. The normal default rule works at this local level, just as it does at the global level, for certain speech acts. For example, if I make a claim, I immediately incur a burden of proof to provide some justification for the proposition asserted by the claim, or I must retract that proposition. However, the requirements for the speech acts of presumption are different. The problem in evaluating any given case is to determine the type of dialogue, and then examine the local argumentation to determine the requirements for each move in the dialogue, where each move fits a type of speech act.

Problems with burden of proof are also very common at the closing stage of the dialogue. At this stage the problem that commonly occurs in connection with an argument from ignorance is called the closed world assumption in AI. This assumption poses the question of when a dialogue can be closed off on the grounds that the search made for information or knowledge during the argumentation stage of the dialogue has been complete enough to prove what is required to be proved. Making such a decision requires not only having set an initial burden of proof at the confrontation and opening stages, but also scanning through the argumentation stage to see if the argument put forward by the one side or the other is strong enough to meet its requirements. It is this question that determines whether closure of the dialogue is appropriate or not. This too is a common issue about burden of proof that can be reasonably disputed at the metalevel of a dialogue.

Rules of evidence in a trial require clear agreement on matters of burden of proof at the outset. For example, if the case is a civil trial, the standard of proof is one of preponderance of the evidence, meaning that the burden is fulfilled by the party who has the stronger argument, on balance. Once it has been agreed at the opening and confrontation stages that the global burden of proof has been set up in this manner, the dialogue can then proceed to the argumentation stage, where both parties present their arguments. But at the local level, during the argumentation stage the burden will shift back and forth, depending on the moves made by each party, as shown by the examples analyzed in Chapter 2.

In Chapter 4, it was shown how a dialogue framework for analyzing burden of proof as a fundamental concept of argumentation theory can be built, and how burden of proof issues can be resolved in this framework. By linking burden of proof to metadialogues in Chapter 6, both subjects of research have been moved forward. Krabbe's demarcation problem of deciding which moves belong to the ground level and which to the metalevel have been solved, at least as far as burden of proof is concerned. The problem of when it is appropriate to shift back and forth between a ground level and a metalevel has been solved by using the device of the three stages of a dialogue and the other tools presented. Even more significantly, the serious problem of the blockage of a dialogue at ground level posed by endless back and forth arguments from ignorance has been solved. Not all burden of proof impasses can be solved, as shown by the political debate example, but the legal examples show that if requirements for burden of proof are properly set at the opening stage, and other procedural requirements are met in the argumentation stage, such an impasse can be broken by moving to a burden of proof metadialogue.

10. An Explanation-Based Approach to Modeling Standards of Proof

In this section we show how, despite the impressive difficulties of defining standards of proof described in Chapter 2, Section 3, recent research in the field of AI and law has provided resources for moving forward to a method of precisely modeling the notion of burden of proof by basing it on standards of proof. This research is building explanation-based models of evidential reasoning that combine argumentation with explanation. Bex and his colleagues (Bex et al., 2010; Bex, 2011,) have built a formal model that enables an explanation to be modeled as a story represented as a script of the kind studied in cognitive science (Schank and Abelson, 1977). This theory is called a hybrid model of evidential reasoning because it combines explanations and arguments in such a manner that an explanation can be supported by pro evidence, and can be attacked by contra evidence. The stories representing the claims put forward by two opposed sides in the case can thus be reasoned about using evidential arguments.

In the hybrid model, an explanation is built around a story that hangs together because it conforms to common knowledge about the way things can be generally expected to work in situations all of us are familiar with in everyday life. As shown in Chapter 4, Section 10, a script is a sequence of actions and events that are connected together in such a way that we can understand it, and also detect parts of it that do not make sense. For example, a script could be my swinging a golf club, hitting the golf ball, the golf ball flying through the air, the golf ball landing on the grass and the golf ball rolling toward the flag but stopping short of it. However, suppose in a different script after I hit the golf ball it zooms a mile into the air, lands on the green and rolls into the cup. This sequence of events and actions is incomprehensible (unless additional information can be brought forward to fill it out, for example, it was part of a science experiment on propulsion of small objects). As the story stands, it does not provide a plausible explanation of how I managed to get a hole in one. An explanation that contains inconsistencies, has large gaps that cannot be filled in, has events in the wrong order or otherwise does not make sense, can be criticized as implausible because it fails to match a comprehensible script. As shown in Chapter 5, Section 9, an explanation can be tested by an examination dialogue.

This kind of plausible reasoning has ten leading characteristics.

1. Plausible reasoning proceeds from premises that are more plausible to a conclusion that was less plausible before the argument.
2. Something is found plausible when hearers have examples in their own minds.
3. Plausible reasoning is based on common knowledge.
4. Plausible reasoning is defeasible.
5. Plausible reasoning is based on the way things generally go in familiar situations.
6. Plausible reasoning can be used to fill in implicit premises in incomplete arguments.
7. Plausible reasoning is commonly based on appearances (perception).
8. Stability is an important characteristic of plausible reasoning.
9. Plausible reasoning can be tested by examination, and by this means confirmed or refuted.
10. Plausible reasoning is closely related to inference to the best explanation.

The characteristics of plausible reasoning in Rescher's theory (Rescher, 1976) show it to have properties comparable to those of the kind of plausible reasoning often described in ancient Greek rhetoric (Tindale, 2010, 81). But plausibility is a property of explanations as well as arguments.

The hybrid theory of Bex and his colleagues uses arguments to support (or attack) explanations by bringing forward evidence that supports (or attacks) the plausibility of an explanation. In inference to the best explanation, multiple explanations are generated, and comparatively evaluated according to criteria that express the degree to which they conform to the evidence and their plausibility. Three criteria are defined using argumentation theory. Arguments based on evidence can be used to show that an explanation is consistent or inconsistent with the evidence. Arguments may also be used to reason about the plausibility of an explanation, as the validity and applicability of causal rules can become the subject of an argumentation process. Arguments about the plausibility of explanations are based on plausible reasoning (Rescher, 1976) of a kind that is comparable to Baconian probability reasoning as described in Chapter 1, Section 10. This kind of evidential reasoning is carried out by using commonsense knowledge about how the world generally works in familiar situations, using scripts of the kind described in Chapter 4, Section 10.

The best way to briefly describe how the hybrid theory can be used to model burden of proof is to use an example from (Bex and Walton, 2012). It is a civil case where the standard of proof is that of preponderance of the evidence, *Anderson v. Griffin* (397 F.3d 515). In this case, the driveshaft suddenly broke on a tractor-trailer proceeding down an interstate highway, severing the connection between the brake pedal and the brakes. Debris kicked up from the surface of the highway (road junk) and struck a pickup truck following the tractor-trailer. The pickup truck crashed into a part of the tractor-trailer and a car following the pickup truck struck the wreckage from the collision between the two trucks, injuring the two people in the car. The plaintiff (the dual party of the two people in the car), sued the truck dealer, because he was held to be responsible for the technical maintenance of the trailer.

The plaintiff's argument was based on the following story. Three weeks earlier, the trucking company who owned the tractor-trailer had noticed a looseness in the driveshaft and had asked the truck dealer to tighten the driveshaft. The dealer tightened all the joints except for the middle one, the one that broke. This explanation was supported by the truck dealer's records stating that the repairmen did not repair that joint. The defendant's story was that the road junk kicked up by the chains hanging from the trailer of the tractor-trailer could have been the cause of the crash. These two stories are represented in Figure 6.8 from (Bex and Walton, 2012, 9). Open arrows denote causal links, closed arrows denote evidential (argument) links and the roundhead arrow denotes evidential contradiction. White boxes

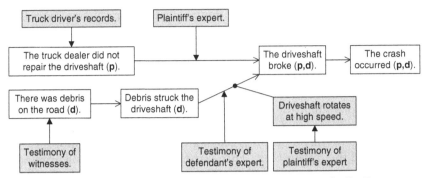

FIGURE 6.8. Two Explanatory Stories in the *Anderson v. Griffin* Case

are part of the sequence of an explanation. The bold letters **p** or **d** denote whether an event is part of the plaintiff's or the defendant's explanation. Grey boxes represent evidence that is part of the argumentation structure of the reasoning.

The plaintiff's story is shown in the sequence of white boxes going from left to right at the upper level of Figure 6.8. The defendant's story is shown just below it. We need to see how the burden of proof shifts from one side to the other. At the beginning, the defendant (the truck dealer) has the tactical burden of proof. If he does not critically question the plaintiffs' explanation or provide an alternative explanation for the crash, the jury will rule for the plaintiff. The defendant gave such an alternative explanation when he claimed that debris struck the driveshaft. Statements made by witnesses support the contention that there was debris on the road. The plaintiff now has the burden of persuasion and production to support their explanation while the defendant at this point only needs to cast sufficient doubt on this explanation. He has already done that by providing a reasonable alternative, which is at least as good as the plaintiffs' explanation. If a verdict were to be given now, the judgment would go against the party with the burden of persuasion, the plaintiffs, because they have failed to meet the burden of production by producing further evidence. Next the plaintiff fulfills their tactical burden of proof by producing evidence: an expert witness is brought forward who states that the crash was caused by the defendant's failure to repair the driveshaft.

Next, the tactical burden shifts to the defendant, as the plaintiffs' extended explanation is now better because it is supported by more evidence. The defendant meets this burden by bringing in new expert testimony stating that the accident had been caused by debris on the highway that was yanked up against the driveline by chains hanging from the truck. The shifting of the burden of proof in the sequence of argumentation is shown in Figure 6.9.

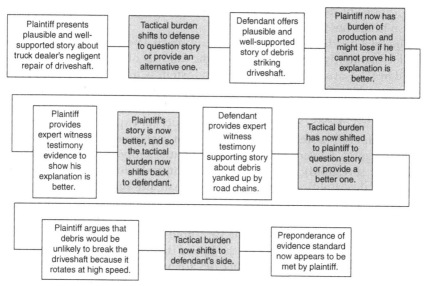

FIGURE 6.9. Shifting of Burden of Proof in *Anderson v. Griffin*

Both explanations now have equal support, so the tactical burden again shifts back to the plaintiffs, who decide to attack the defendant's explanation. Next, the plaintiff argued that a piece of road junk would be highly unlikely to strike the driveshaft with enough force to break it, given that the driveshaft rotates at such a high speed (twenty-seven times a second). At this point the plaintiffs' explanation is slightly better than the defendant's and if so it can be inferred that the preponderance of evidence standard has been met. As it happened in this particular case however, in the case the jury ruled in favor of the defendant. Perhaps they may have found that the plaintiffs' argument from expert opinion was not strong enough.

The stages in this sequence of evidential reasoning as the burden of proof has shifted from side to side in this case are summarized in Figure 6.9. Figure 6.9 shows how the tactical burden of proof shifts back and forth during a trial as each side offers arguments to support its own explanation and arguments to weaken the explanation offered by the other side.

To model burden of proof in a case, the hybrid theory requires a set of evidential rules, a knowledge base composed of a consistent set of statements comprising the evidence and a set of commonsense assumptions based on common knowledge and scripts. The argumentation part of the hybrid theory is similar to the ASPIC+ logic, the rule-based system used by Prakken and Sartor to model burden of proof. Evidential arguments are built by taking evidence and assumptions as premises and chaining applications of defeasible *modus ponens* into tree-structured arguments as shown in the Los Gatos example.

Finally, let's show how the hybrid approach (Bex and Walton, 2012) models standards of proof based on the analysis of these standards provided by the Carneades Argumentation System (Chapter 2, Section 3). To show how this works, let us examine how the model defines the standards of clear and convincing evidence and beyond reasonable doubt. To meet the standard of clear and convincing evidence an explanation E needs to be good in itself as well as much better than each competing explanation E'. To be good, E needs to have high evidential support ($es(E) > \alpha$, where α is some threshold) and low evidential contradiction ($ec(S) < \beta$, where β is some threshold). To be much better than any alternative E', E needs to have either significantly higher evidential support ($es(S) - es(S') > \gamma$, where γ is some threshold) or significantly lower evidential contradiction ($ec(E') - ec(E) > \delta$, where δ is some threshold).

Two detailed examples of how the hybrid theory models the beyond reasonable doubt standard in criminal cases are presented in (Bex and Walton, 2012). An explanation meets the beyond a reasonable doubt standard if it is strong, and it is much stronger if there are no plausible competing explanations. How the beyond reasonable doubt standard is modeled by the hybrid theory can be roughly summarized as follows. An explanation of the facts of a case meets the beyond reasonable doubt standard only if it meets two requirements. First, it must be a highly plausible explanation in itself. Second, the competing explanations (if there are any) must be implausible. The first criterion requires that the explanation that was given provides a coherent story that stands up to criticism and is highly plausible, is strongly supported by the evidence up to a high threshold and meets other criteria of a good explanation. The second criterion requires that each of the competing explanations that has been offered is so weak that it fails to raise a reasonable doubt.

So far this explanation-based approach to modeling standards of proof is new and has not been extensively tested or developed. We present it here to show that there is a least a hypothesis for working toward building a computational model of standards of proof that shows promise of moving forward to implementing Allen's notion that inference to the best explanation needs to play a larger role in evidential reasoning. This hypothesis could be a way of responding to the grave difficulties in gaining precise and useful account of burdens and standards of proof.

7

Burdens of Proof in Different Types of Dialogue

Most of the literature on burden of proof in argumentation studies and AI has concentrated so far on the persuasion type of dialogue. This concentration is natural enough, because the bulk of this literature has concentrated on burden of proof in legal argumentation. The most significant exception is deliberation dialogue, where some recent work has begun to tentatively investigate burden of proof in that setting. The problem now posed is whether burden of proof operates in deliberation dialogue in the same way that it operates in persuasion dialogue, or whether there are essential differences in this regard between the two types of dialogue.

This chapter analyzes four examples of deliberation dialogue where burden of proof poses a problem. Based on analysis of the argumentation in these examples, a working hypothesis is put forward. It is that burden of proof only becomes relevant when deliberation dialogue shifts, at the beginning of the argumentation stage, to a persuasion dialogue. The hypothesis is that the shift can be classified as embedding one type of dialogue into another, meaning that the goal of the first type of dialogue continues to be supported once the transition to the second type of dialogue has been made (Walton and Krabbe, 1995, 102). In other instances, it is well known that a shift can be illicit, where the advent of the second dialogue interferes with the fulfillment of the goal of the first one. It has also been shown that such shifts can be associated with fallacies, as well as other logical and communicative problems (Walton, 2007, chapter 6).

The work in this chapter extends existing formal models of deliberation dialogue to analyze four examples of deliberation dialogue where burden of proof is at issue or poses an interesting problem. The examples are used to show (1) that the eight stages in the formal model of McBurney, Hitchcock and Parsons (2007) need to be divided into three more general stages: an opening stage, an argumentation stage and a closing stage; (2) that deliberation dialogue typically shifts to persuasion dialogue during the argumentation stage and (3) that burden of proof is only operative during the argumentation stage. What is shown in general is that deliberation is, in the typical type of case, a mixed dialogue in which there is a shift to persuasion dialogue in the middle. These investigations suggest a conclusion that many would hold to be contrary to the commonly accepted climate of opinion about burden of proof in deliberation, which holds that burden of proof is an important notion for evidential reasoning in deliberation, especially in matters of public safety where danger is involved in such a manner that would require setting a high burden of proof against any contemplated course of action that might be dangerous.

These conclusions are likely to remain controversial for the next little while, as they have not yet been widely discussed in the argumentation literature, and because they contravene the widely accepted view that burden of proof plays an important role in deliberation. However, in this chapter we are moving into territory that has not been thoroughly investigated in the literature on logical argumentation, and so it is best to see the chapter as posing a number of unsolved problems that arise within the methodology of argumentation studies. The recommendations made in this chapter are tentative, and merely represent an approach that may help us move forward to extend the notion of burden of proof in persuasion dialogue analyzed in the previous chapters. In particular, more research needs to be carried out on dialectical shifts of the kind studied in Walton and Krabbe (1995) before we can grasp precisely how persuasion dialogue is embedded in deliberation dialogue in such a manner that the persuasion dialogue can help the deliberations move forward in a constructive manner.

1. Some Examples

Deliberation is different from persuasion dialogue. In persuasion dialogue, there is some claim at issue, and the object of the dialogue is to prove or to disprove that claim. Deliberation has a different kind of goal. It is to solve a problem about what course of action to take. The problem statement is not a proposition, but a question, called a governing question by McBurney, Hitchcock and Parsons (2007). Examples of these are: "Where should we go to dinner?" and "How can we provide all Americans with health care insurance?." The goal of a deliberation is to find a solution to a common problem. Unlike persuasion dialogue, there are no winners and losers. Everyone wins if the dialogue is successful. Does burden of proof have a place in this type of dialogue? It seems so because arguments go back and forth in a deliberation dialogue, and once an argument is brought forward, like "Ricardo's is the best place to go for dinner, because their food is organic," it requires evidence to back it up if it is challenged. It appears then that understanding how burden of proof works in it is an important step in the study of deliberation dialogue as a form of group decision making.

Deliberation dialogue begins with a problem, and the goal of the dialogue is to find a solution to the problem, usually some action to take, typically in a collaborative, not an adversarial context. This process often involves a brainstorming phase in which ideas are put forward that are not yet formulated as proposals that the person who put the idea forward has a commitment to defend. Arguments for and against these ideas can be gathered, with every party providing pro and con arguments for all the alternatives on the table. During this brainstorming phase, parties will put forward con as well as pro arguments for the ideas put forward by other parties. Only later in the deliberation, after the brainstorming phase, do parties propose and defend specific solutions. It is during this phase, as we will contend later, that the deliberation dialogue shifts to a persuasion dialogue.

The first example was a debate in the Rhode Island Assembly on whether or not to bring in no-fault insurance, fully described in Lascher (1999), and cited in more abbreviated form as an example of deliberation dialogue in Walton (1998, 169). One side proposed bringing in a new system of no-fault insurance in Rhode Island, arguing that insurance rates were too high, and that paying the premiums had become burdensome. The goal of both sides was presumably to lower insurance rates, if possible. The opposed side argued that the proposed no-fault system would unfairly make good drivers pay for bad drivers, and would fail to lower insurance premiums.

This example initially appears to be one of a deliberation dialogue in which two groups engaged in discussion with each other are arguing from what they take to be their common commitments. The point of disagreement is that each side is doubtful that the proposals for action put forward by the other side will fulfill the central goal both agree on. This case looks like deliberation because there were two sides, for and against, and each used practical reasoning to support its side, often by employing argumentation from consequences. The no-fault side argued, for example, that the change to no-fault insurance would reduce costs of coverage. The opposed side argued, for example, that no-fault unfairly makes good drivers pay for bad drivers. In this case, each side put forward some general or global action that it advocated. The no-fault side advocated changing to a no-fault system. The opposed side argued for retaining the status quo.

The second example treated here is Wigmore's purse case comparable to his property investment case described in Chapter 1, Section 3. In this example, A, arrives at his destination and steps out of his car to the crowded sidewalk, sees a purse lying there, picks it up and looks around to see who may have dropped it. In the example, it is supposed that M steps up to him, and claims the purse as his own. At first A is in doubt; hence, inaction as to surrendering it. Then he says to M, "Prove your ownership." In the example, it is supposed that M makes a statement that is unconvincing, and that A is still in doubt. Hence, A takes no action. But next it is supposed that M describes the contents of the purse exactly. Then "conviction comes to A, and he hands the purse to *M*" (Wigmore, 1935, 440). The argumentation in this case is based on a practical need to take action, and therefore it appears reasonable to classify it as a case of deliberation dialogue.

In this example, *A* does not act on the basis of any legal notion or theory of burden of proof, according to Wigmore's analysis. *A*'s decision is an instinctive one of requiring *M* to remove his doubt before he hands over the purse. As long as *A*'s doubt remains in place, *M* does not get the purse. According to Wigmore (1935, 439), doubt and conviction are the two contrasting states of mind of a person who is confronted with a choice of actions. Doubt leads to inaction, whereas conviction leads to action.

The third example concerns a problem that has arisen recently concerning the importation of active pharmaceutical ingredients from overseas. One example cited concerned imported heparin[1] that was contaminated and that claimed the lives of patients taking pharmaceuticals in which this drug was an ingredient. An energy and commerce committee asked Congress to grant it powers to order recalls of drug products, to block suspicious imports from gaining access to the United States and to require foreign firms to divulge data in inspections. One committee member expressed the problem by saying that according to current practice, the Federal Drug Administration (FDA) must show at the border that imported active pharmaceutical ingredients are unsafe. Instead of the burden being on the FDA to prove that the shipment is unsafe, he suggested, it would be better if the company importing the shipment had the obligation to prove that it is safe.[2] How could this case be analyzed as an instance of deliberation dialogue in which there is argumentation on two sides of an issue and burden of proof is involved? Finally, there is a fourth example that needs to be examined fully because it is especially controversial and problematic.

The precautionary principle was introduced in Europe in the 1970s to give the environmental risk managers regulatory authority to stop environmental contamination without waiting for conclusive scientific evidence of harm to the environment. It is controversial how the principle should be defined, but a rough definition that provides a beginning point for discussion is: if an action or policy might cause severe or irreversible harm to the public or the environment, in the absence of conclusive scientific evidence that harm would not ensue, the burden of proof falls on the side that advocates taking the action. Note that this definition links the precautionary principle to the notion of burden of proof. It is meant to be applied to the formation of environmental policy in cases like massive deforestation and

---

[1] Heparin is a highly sulfated glycosaminoglycan widely used as an anticoagulant.

[2] This example is a paraphrase of a case described in Joseph Woodcock, "Burden of Proof of Safety Must Fall on Drug Manufacturers," Validation Times, May, 2008, 1–7. Found Dec. 22, 2008 at http://findarticles.com/p/mi_hb5677/is_5_10/ai_n29445559.

mitigation of global warming where the burden of proof is ruled to lie with the advocate.

An early application of the principle was to prohibit the purging of ship bilge contents into the oceans (Freestone and Hey, 1996). Because of lack of scientific data on the effects of the purging of bilge contents on the oceans, scientific risk assessment of the practice was not possible. The application of the precautionary principle gave regulatory officials the authority to prohibit the practice without waiting for scientific evidence that could prove harmful to the environment.

Among criticisms of the precautionary principle is the argument that its application could create an impossible burden of proof for marketing new food products or ingredients (Hathcock, 2000, 225). According to this criticism, excessive precaution can lead to paralysis of action resulting from unjustified fear. Some examples cited include the outbreak of cholera resulting from fear of chlorinated water, and the reluctance to permit food fortification with folic acid to reduce the incidence of birth defects for fear of masking vitamin B12 deficiency (Hathcock, 2000, 255). What is especially interesting is that both defenses and criticisms of the precautionary principle link it closely to the concept of burden of proof.

The precautionary principle was adopted by the U.N. General Assembly in 1982, and was implemented in an international treaty by the Montreal Protocol in 1987. According to the Rio Declaration of 1992, "where there are threats of serious or irreversible damage, lack of full scientific certainty shall not be used as a reason for postponing cost-effective measures to prevent environmental degradation." In some countries, like the United States, the precautionary principle is designated as an approach rather than a principle, meaning that it does not have legal status. In other countries, and in the European Union, it has the legal status of a principle, meaning that is it is compulsory for a court to make rulings in cases by applying it (Recuerda, 2008).

Critics have argued that the precautionary principle can be used to stop the use of any new food products because safety cannot be proved with certainty in any case of a new product (Hathcock, 2000, 258). There is also the problem of judging how serious a harm has to be and how likely it is before the principle can be applied. The principle was originally meant to give regulatory authority to stop environmental contamination, but once made into law, as Bailey (1999, 3) pointed out, it could conceivably be applied to all kinds of activities. Applying the principle to other areas, for example, inventors would have to prove that their inventions would never do harm before they could be marketed to the public (Bailey, 1999, 4).

One of the problems with implementing the precautionary principle is that there are open questions about the standard of proof that should be applied to the side advocating the action or policy question. It would seem that, because the principle is supposed to be applied under conditions of lack of full scientific certainty, a high standard of certainty, like beyond reasonable doubt, would not be appropriate. On the other hand, there are the questions of how serious and widespread the harm needs to be, and how it can be shown that it is irreversible, before the principle should be applied. There is also the question of how it should be judged and how much evidence should be given by the advocate of the action to match the perceived seriousness and likelihood of the harm. The principle needs standards of proof for both sides, but the standards of proof that should be required might vary from case to case.

2. The Formal Structure of Deliberation Dialogue

Deliberation always begins with the formulation of a problem about which action to take in a given set of circumstances. The problem is formulated in a governing question of the kind, "What should we do now?" The first stage of the dialogue comprises both the formulation of the governing question and the circulation of the information about the given circumstances of the decision to be made among all the members of the group. Knowledge of the circumstances is being updated continually and circulated during a typical deliberation dialogue, but the collection of data is typically limited by costs, and in particular by the cost of delaying arriving at a decision on what to do. There is always a tradeoff between arriving at a timely decision on what to do and the improvement of the deliberation that would be made by collecting more relevant information about the circumstances. This opening stage comprises the first two stages represented in the formal model of deliberation dialogue of McBurney, Hitchcock and Parsons (2007, 100) with its eight stages called *open, inform, propose, consider, revise, recommend, confirm* and *close.*

> Open: A governing question, like "Where shall we go for dinner this evening?," expressing a need to take action in a given set of circumstances, is raised.
> Inform: This stage includes information about facts, goals, values, constraints on possible actions and evaluation criteria for proposals.
> Propose: Proposals cite possible action options relevant to the governing question.
> Consider: This stage concerns examining arguments for and against proposals.
> Revise: Goals, constraints, perspectives and action options can be revised in light of information coming in and arguments for and against proposals.
> Recommend: Based on information and arguments, proposals are recommended for acceptance or nonacceptance by each participant.

Confirm: The participants confirm acceptance of the optimal proposal according to some procedure. All participants must do so before the dialogue terminates.

Close: Termination of the dialogue once the optimal proposal has been confirmed.

An important property of deliberation dialogue (McBurney, Hitchcock and Parsons, 2007, 98) is that a proposal may be optimal for the deliberating group but suboptimal for any individual participant. Another feature is that in a proper deliberation dialogue each participant must share his/her individual goals and interests, as well as information about the given circumstances. The goal of deliberation dialogue is for the participants to decide collectively on what the optimal course of action is for the group.

It is important to note that the temporal progress of a real deliberation is not the same as the normative model of the argumentation that should ideally take place in it. The bringing in of information is not restricted only to the opening stage in real instances.

Deliberation needs to proceed under conditions of uncertainty and lack of knowledge about a complex situation that is constantly changing. For this reason, information about the changing situation needs to be updated continually. An important skill of deliberation is to adapt an initial plan of action to new information that comes in reporting changes in the existing circumstances. There is typically feedback in which the agents who are involved in the deliberation may see the consequences of the actions they have already carried out, and need to modify their plans and proposals by updating in light of the new information. For this reason, deliberation dialogue needs to be seen as having an information-seeking dialogue embedded into it. It is constantly shifting from looking at the arguments for and against a proposal and taking into account the new information about the changing factual circumstances of the case being considered. At the opening stage, the *inform* function is employed to collect a database of information concerning the circumstances of the given situation, but later additions and deletions to it need to be made during the argumentation stage.

The opening stage also has a brainstorming phase in which ideas are put forward, but not yet as firm proposals that the participant who voiced the proposal is committed to defending. Nor is he committed to attacking opposed proposals at this point. At this stage, a participant may bring out weak points in a proposal he has articulated, and find strong points in a proposal someone else has voiced. But then at the *revise* phase, there is a shift. At this point, when a party puts forward a proposal, he is advocating it as the best solution to the problem posed in the opening stage. Thus, at this point, we are no longer in the opening stage. We are now in the argumentation stage. The argumentation stage also includes the *recommend* phase, but the last two phases in the McBurney, Hitchcock and Parsons model, the *confirm* and *close* phases, are parts of the closing stage of the deliberation dialogue.

Now we have divided the eight phases of the McBurney, Hitchcock and Parsons model into three more general stages, and there is a problem that arises. In the middle stage, the argumentation stage, each party defends the proposal he or she has advocated as solving the problem set at the opening stage, and attacks the alternative proposals put forward by other parties. In this stage, there has been a shift to a persuasion dialogue, even though later on, at the closing stage, the discussion will shift back to a deliberation dialogue. Now there is a problem of how to track such a shift in a given case, and to approach this problem we need to be clearer on how to distinguish deliberation dialogue from persuasion dialogue in certain kinds of cases.

3. Deliberation versus Persuasion Over Action

The characteristics of persuasion dialogue and deliberation dialogue have now been made clearly enough in general outline. In a deliberation dialogue, the central goal is for the participants to arrive at a decision on what to do, given the need to take action in a set of circumstances requiring a choice. In deliberation dialogue, the problem of what to do is discussed among the participants, and then some proposals for action emerge from the discussion. Making a proposal is defined as a kind of speech act (Walton, 2006) where a party suggests a course of action as providing an answer to the question posed at the opening stage of the dialogue. In the model of McBurney, Hitchcock and Parsons, the need to take action is expressed in the form of a governing question like, "How should we respond to the prospect of global warming?" It is a governing question because it is set at the opening stage, governs the moves in the argumentation stage and determines which proposal should be accepted at the closing stage. In a persuasion dialogue, the proponent's goal is to prove the proposition that is designated at the opening stage as her ultimate thesis (to a standard of proof set at the opening stage) by means of a chain of argumentation. The goal is one of rational persuasion by offering reasons for the other party to come to accept some statement he initially doubted.

This distinction between the two types of dialogue seems clear enough in outline but there is a commonly recurring problem on the issue of whether the discussion should be classified as a persuasion dialogue or a deliberation dialogue. This problem arises in cases of persuasion over action, referring to cases in which one party is trying to persuade another to take a particular course of action. Consider, for example some topics of debates taken from Debatepedia.[3]

---

[3] http://debatepedia.idebate.org/en/index.php/Past_Debate_Digest_topics (accessed Nov.18, 2011).

- Should there be a ban on sales of violent video games to minors?
- Should there be mandatory ultrasounds before abortions?
- Should colleges ban fraternities?
- Should public schools be allowed to teach creationism alongside evolution as part of their science curriculum?
- Should governments legalize all drugs?
- Should illegal immigrants in the United States be allowed to obtain drivers licenses?

In each case, the topic of the debate concerns a decision to take action, suggesting that the debate should be classified as a deliberation dialogue rather than a persuasion dialogue. However, note that in each case the debaters are not themselves actually making the decision of what to do in the case they are discussing. Instead, they are putting forward and examining the arguments on both sides in order to arrive at some conclusion on what would generally be the best thing to do. For example, the debaters on Debatepedia discussing the issue of whether colleges should ban fraternities are not actually making the decision to ban fraternities, nor are they in a position to take action to ban fraternities in all colleges, even if their debate may come to the conclusion that all colleges should ban fraternities. Thus, it is improper, on the criteria given in Section 2, to classify these debates as deliberations. They need to be classified as persuasion dialogues over action. Looking at this list of typical debates, and many other debates of the same kind that can be found on Debatepedia, the conclusion can be drawn on the basis of this sample that persuasion dialogues over action of this kind are common.

Some recent research in AI (Atkinson et al., 2013) provides a solution to this problem by studying an everyday example where a group of participants at a conference has to make a decision on which restaurant to go to at the end of the day. They are deciding between three restaurants, and during the dialogue each of them makes proposals, each makes his or her preferences clear to the others and they bring forward arguments on which restaurant is best based on the factors that are important in their preferences. Through the study of this example, where the discussion is modeled computationally using Prolog, four general points of contrast are drawn between persuasion dialogue and deliberation dialogue.

1. Persuasion starts with a conflict, and therefore it has a highly adversarial structure, whereas in a deliberation dialogue, the participant taking an initiative is not trying to prove anything, but to find an acceptable solution to the problem that everybody can agree on as the best decision.
2. The roles in a persuasion dialogue can be asymmetrical, whereas in a deliberation dialogue each participant has essentially the same role as each other participant.

3. In a persuasion dialogue each participant has internal commitments that taken together represent his or her personal and private point of view. In order to be persuaded, it is necessary for the persuading party to get the other party to alter this point of view and change his/her mind concerning the issue being discussed.

4. In a persuasion dialogue the party doing the persuading continually has to supply arguments that present information that can be used to convince the other party, whereas in a deliberation dialogue all the parties supply and request information.

These differences are important, but are not by themselves sufficient to enable the distinction to be drawn in a particular case whether a discussion should be properly classified as a deliberation dialogue or a case of persuasion over action.

In the analysis of the example presented in (Atkinson et al., 2013) the key distinguishing factor is the set conditions for the use of the speech acts in each type of dialogue. Although participants use similar kinds of speech acts in both types of dialogue, like making assertions, asking questions and so forth, there are differences in the two dialogue types between what the participants are supposed to be doing by using these performatives and what the other participants can conclude from them. The difference resides in the criteria used to justify the choice made by the participants at the conclusion of the dialogue. In a persuasion dialogue, these criteria are determined on the basis of individual preferences by the party who is to be persuaded. In a deliberation dialogue, these criteria are not yet fixed at the opening stage and are formed during the course of the dialogue.

When it comes to the closing stage in the decision on what to do in a deliberation dialogue, only one set of criteria representing the preference of the group as a whole is used. In a persuasion dialogue, the underlying purpose the participant has in mind is to use the personal criteria of the party being persuaded to get him to accept the view he formerly opposed. In a deliberation dialogue, any participant needs to use the criteria of all the participants to move forward to form a decision that is acceptable to the whole group. It is essentially for these reasons that there is no burden of proof in a deliberation dialogue, unlike the case of the persuasion dialogue, where the proponent is required to satisfy the criteria to which the other party is committed. The ultimate aim in the persuasion dialogue is for the proponent to prove something to the other party by making certain sorts of moves, especially by bringing forward arguments based on the other party's commitments, arguments that once they are all connected together, form a network of proof that is strong enough to meet the standard of proof required.

The goal of argumentation used in a deliberation is not to attempt to persuade the other parties to become committed to one's own proposal. However, after one's own proposal has been articulated to the other participants, they will expect the proposer to defend it by presenting arguments and information to support it. The offering of the supportive evidence takes place within a persuasion dialogue that is embedded into the deliberation dialogue. To make this embedding possible, there has to be a dialectical shift (Walton and Krabbe, 1995, 100–116) to persuasion dialogue in order for reasons for or against the course of action being recommended in the proposal to be supported and criticized. In some instances, in order to analyze and evaluate the argument, it doesn't matter whether the argument is part of a deliberation dialogue or a persuasion dialogue. However, in other instances it makes a great deal of difference what type of dialogue we see the argument as part of. In some cases, the issue of whether the argument can be fairly criticized as committing a fallacy depends on this contextual issue of whether it should be seen as part of a deliberation dialogue or a persuasion dialogue.

Argument from consequences has been cited as a fallacy in *Introduction to Logic* (Rescher, 1964, 82). In this textbook, the reader is warned that "logically speaking," it can be "entirely irrelevant that certain undesirable consequences might derive from the rejection of a thesis, or certain benefits accrue from its acceptance." The following example is cited as a case in point: "The United States had justice on its side in waging the Mexican war of 1848. To question this is unpatriotic, and would give comfort to our enemies by promoting the cause of defeatism." According to the analysis of this example presented by Rescher, the argument from consequences can be classified as a fallacy of relevance. The argument that questioning that the United States had justice on its side in this war would give comfort to our enemies is irrelevant. The reason is that the proposition to be proved in the discussion is whether the United States had justice on its side in the Mexican war of 1848. The issue in the discussion is an ethical and political one about right and wrong in international affairs. In this particular discussion, the statement that questioning whether the United States had justice on its side in this war would give comfort to our enemies, while it is a factual statement that might well be true, it is not relevant evidence for accepting or rejecting the conclusion issue.

This analysis of the Mexican war example as being an instance of fallacy of irrelevant argumentation can be brought out more clearly by showing how part of the argument fits the argumentation scheme for argument from negative consequences. The two argumentation schemes representing arguments from consequences (Walton, 1996, 75) can be used for this purpose. First, there is the scheme for argument from positive consequences, where A is a proposition representing a state of affairs that can be an outcome of an action.

Argument from consequences can take either one of the two following forms. The first is called Argument from Positive Consequences.

Premise: If *A* is brought about, good consequences will occur.
Conclusion: Therefore *A* should be brought about.

The second is called argument from negative consequences.

Premise: If *A* is brought about, then bad consequences will occur.
Conclusion: Therefore *A* should not be brought about.

Three critical questions can be used (Walton, 1996, 76–77) to provide resources in a dialogue for a respondent to express skeptical doubts about an argument having this form.

CQ1. How strong is the probability or plausibility that these cited consequences will (may, might, must) occur?
CQ2. What evidence, if any, supported the claim that these consequences will (may, might, must) occur if *A* is brought about?
CQ3. Are there consequences of the opposite value that ought to be taken into account?

An instance of an argument from positive or negative consequences can be perfectly reasonable in some cases, even though such an argument is defeasible and inherently open to critical questioning that may cause it to default. The argument can be strong (defeasibly) if positive or negative consequences are cited as reasons to support the proposed course of action. It can be weak if it fails to address an appropriate critical question. Or it can even be fallacious if it is used in some way to block or interfere with the proper course of a dialogue.

To see how the argument in this particular instance can rightly be accused of committing a fallacy of relevance, it helps to diagram the structure of the argument using the argument mapping tool of the Carneades Argumentation System (see Figure 7.1).

The proposition at the bottom right stating that giving comfort to our enemies is a bad consequence has been inserted as an implicit premise. Taken together with the other premise just above it, the argument from negative consequences is shown leading to the conclusion shown on the left. How the scheme for argument from negative consequences joins these two premises together to form the inference to the conclusion shown at the left is indicated by the node containing the name of the scheme for argument from consequences. It is represented as a contra argument, because it is a negative form of argument from consequences.

Both sets of premises offer reasons why we should not question the assertion made in the conclusion of the argument that the United States had justice on its side in the Mexican war. But let's take a closer look at the part of the argumentation forming the argument from negative

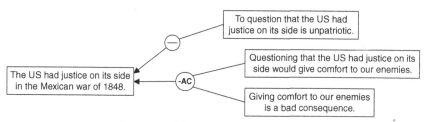

FIGURE 7.1. Argument Diagram of the Mexican War Argument

consequences. It might be true that giving comfort to our enemies is a bad consequence, and it might be true that questioning whether the United States had justice on its side would give comfort to our enemies. However, although both premises might well be true, the argument itself is not relevant. Actually, it is not relevant, but it does somehow look like it is relevant. An explanation for this appearance can be given as follows: it would be relevant if the dialogue were that of a deliberation on how to act in this situation, for giving comfort to enemies could be a bad consequence, and one that what we would justifiably want to avoid when making a decision about what to do, especially in the context of a war or other dangerous situations where loss of life could be involved. But would the very same argument from consequences shown in Figure 7.1 be relevant in a persuasion dialogue about whether Mexico or the United States had justice on its side of the kind one might have in a history class or an ethical discussion, in which the issue is what country had justice on its side in this particular war? Arguing from negative consequences, by saying that questioning that the United States was in the right would promote defeatism and have bad consequences for the national interest, would not be relevant evidence in such a persuasion dialogue about right and wrong. Why would such an argument from negative consequences ever seem to be relevant? One explanation is that there has been a shift to a different issue, and thus the failure of relevance of the argument from negative consequences is a reason for criticizing the argument on the basis that a fallacy of irrelevance has been committed by it. The problem is that the argument from negative consequences, although not an inherently unreasonable form of argument in itself, in the particular setting of the persuasion dialogue about right and wrong in the Mexican war, fails to fulfill its burden of persuasion.

There is another problem generally with analyzing this argument because it is an everyday conversation argument, and there was no judge or debate moderator who set the ultimate issue to be proved at the opening stage of the persuasion dialogue. Still, however, we can say that the argument from negative consequences has been used in a fallacious way in this instance to commit a fallacy of irrelevance on the basis that the way the conclusion is stated provides evidence to support the hypothesis that the discussion is a persuasion dialogue and that its ultimate conclusion is the statement that the United States had justice on its side in waging the Mexican war of 1848.

Reconsidering these subtle but important differences between persuasion dialogue and deliberation dialogue in light of the examples outlined in Section 1 will help to throw further light on how burden of proof is involved.

4. Analysis of the No-Fault Insurance Example

There are two basic types of persuasion dialogue, depending on how the burden of proof is allocated (Walton and Krabbe, 1995). In a dispute (symmetrical persuasion dialogue) each side has a thesis to be proved. For example, White (a theist) has to prove that God exists while Black (an atheist) has to prove that God does not exist. In a dissent, one party has a thesis to be proved while the other, in order to win, needs only to cast doubt on the first party's attempts to prove her thesis, so that her burden of proof is not met. For example, White (a theist) has to prove that God exists while Black (an agnostic) needs only to cast doubt on White's attempt to prove her thesis, so that her burden of proof is not met. Thus, the following propositions follow:

- In a dispute, both sides have a burden of proof. One side has to prove *A* and the other has to prove not *A*.
- In a dissent, one side has to prove *A* while the other only needs to cast doubt on the attempts of the first side to prove *A*.
- It follows that the standard of proof needed to win must be set at the opening stage.
- In persuasion dialogue, burden of proof must be set at the opening stage.

At first sight, the way the burden of proof needs to be organized in the no-fault insurance example seems comparable to a persuasion dialogue.

To see whether it is, let us examine some features of the no-fault insurance example. In this example, the burden of proof seems initially to be set in a clear way that is unproblematic. Each side has a proposal. The proposal of the one side is the opposite of that of the other. This suggests a dispute about what action to take. One side proposed bringing in a new system of no-fault insurance, while the opposed side was against the no-fault system. This case shows how serious problems of burden of proof can arise during the argumentation stage. Consider the example dialogue in Table 7.1.

From such examples, we can see that the speech act of making a proposal is very much like the speech act of putting forward an argument in a persuasion

TABLE 7.1. *Argumentation in Dialogue Format in the No-Fault Insurance Example*

	No-Fault Side	Opposed Side
1	I propose a no fault-system.	On what grounds?
2	The insurance rates are too high under the existing system.	How can you prove that a no-fault system would lower the rates?
3	How can you prove that a no-fault system would not lower the rates?	It's up to you to prove that a no-fault system would lower the rates.
4	No, it's not.	Yes, it is.
5	You made the claim that a no-fault system would not lower the rates.	No I didn't. Where did I say that?
6	Your argument depends on that claim.	Not really, I just know that the rates are too high under the existing system.
7	Unless you can prove that a no-fault system would not lower the rates, your argument fails.	No, you need to prove that a no-fault system will lower the rates.
8	OK, but my reason is that it would lower the rates.	Well then, prove that this claim is not true.

dialogue, and involves the same problems arising from disputes about burden of proof. The proposal itself can be seen as a claim put forward, with a local burden of proof comparable to that attached to the speech act of putting forward an argument in a persuasion dialogue.

The making of a proposal advocates a proposition for action that needs to be supported, if questioned or attacked, by putting forward other propositions that are offered as reasons in favor of accepting the proposal. On the analysis advocated here, these other propositions are linked to the proposition that is the proposal by practical reasoning, including related schemes like argumentation from consequences. Both sides share the common goal of lowering the insurance rates if possible, but the disagreement is about the best way to carry out the goal. One side has put forward a proposal to bring in a new system of no-fault insurance, while the other side argues against this proposal. We are not told whether the other side has a different proposal of its own to put forward. It may be that they have no new proposal and are simply arguing for sticking with the old system until a better one can be found, or perhaps for modifying the old system in some way.

What can we say about the role of burden of proof in such a case? In the way the cases are described earlier, it would appear that the side who has proposed bringing in the new system of no-fault insurance would have to make a strong enough case for their proposal to show that it is significantly better than the alternative of sticking with the old system. For example, if they put forward a series of arguments showing that the new proposal was only marginally better than the existing system, that might not be regarded as a sufficient reason for making the change to the new system, or regarding it as worth doing. To convince the audience that the new proposal is the best way to move forward in reducing insurance rates, they would have to provide sufficient reasons to show that the new system has advantages over the old system that warrant the cost of making the change. But this conservatism is just another argument from negative consequences (the negative consequence of added costs).

Does each side have a burden of proof to fulfill, set at the opening stage of the deliberation dialogue, or can a side win the dialogue merely by proving that its proposal is stronger than all the alternative ones, even if it is only slightly stronger? Some might say that this question depends on how the burden of proof was set at the opening stage of the deliberation dialogue. Was the deliberation set up in such a way that only the no-fault side has a positive burden to prove its proposal is acceptable, while the opposed side can be allowed not to prove any proposal that it has advocated?

However, a different answer to the question can be given. The answer is that in a deliberation dialogue, proposals are put forward only during the argumentation stage. If this is right, burden of proof is set and is operative only during the argumentation stage. If this is so, the question is raised whether burden of proof only comes into play during the argumentation stage. The next question raised is whether the argumentation stage consists of a persuasion dialogue. Only when proposals are put forward, during the argumentation stage, does burden of proof come into play. If this approach is right, it suggests that the deliberation has shifted to a persuasion interval during the argumentation stage. These questions can be investigated by taking a closer look at the argumentation used during the argumentation stage of the no-fault insurance example.

Much of the argumentation in the no-fault insurance example fits the argumentation schemes for practical reasoning and argument from consequences (highly characteristic of deliberation). The argumentation scheme in such a case is the one for practical reasoning (Atkinson, Bench-Capon and McBurney, 2006). The simplest form of practical reasoning, called instrumental practical reasoning, is represented by the following scheme (Walton, Reed and Macagno, 2008, 323):

Major Premise: I (an agent) have a goal G.
Minor Premise: Carrying out this action A is a means to realize G.

Conclusion: Therefore, I ought (practically speaking) to carry out this action *A*.

Below is the set of critical questions matching the scheme for instrumental practical reasoning (Walton, Reed and Macagno, 2008, 323).

CQ₁: What other goals do I have that should be considered that might conflict with *G*?

CQ₂: What alternative actions to my bringing about *A* that would also bring about *G* should be considered?

CQ₃: Between bringing about *A* and these alternative actions, which is arguably the most efficient?

CQ₄: What grounds are there for arguing that it is practically possible for me to bring about *A*?

CQ₅: What consequences of my bringing about *A* should also be taken into account?

The last critical question is very often called the side effects question. It concerns potential negative consequences of a proposed course of actions. Just asking about consequences of a course of action being contemplated could be enough to cast an argument based on practical reasoning into doubt.

The basic scheme for practical reasoning is instrumental, but a value-based scheme has been formulated by Atkinson, Bench-Capon and McBurney (2005, 2–3):

- In the current circumstances *R*
- we should perform action *A*
- to achieve new circumstances *S*
- which will realize some goal *G*
- which will promote some value *V*.

According to this way of defining the scheme, values are seen as reasons that can support goals. The scheme for value-based practical reasoning can be classified as a composite of instrumental practical reasoning and argument from values.

In the account of schemes given in (Walton, Macagno and Reed, 2008), argument from values is seen as a distinct type of argument in its own right, with two species. The first species is called argument from positive value.

Premise 1: Value *V* is *positive* as judged by agent *A* (value judgment).

Premise 2: The fact that value *V* is *positive* affects the interpretation and therefore the evaluation of goal *G* of agent *A* (If value *V* is *good*, it supports commitment to goal *G*).

Conclusion: *V* is a reason for retaining commitment to goal *G*.

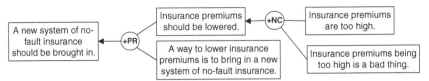

FIGURE 7.2. Practical Reasoning and Argument from Consequences in the No-Fault Example

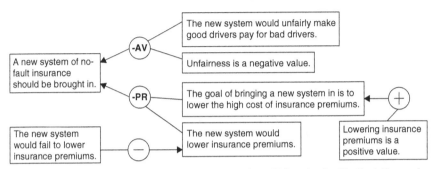

FIGURE 7.3. Practical Reasoning and Argument from Values in the No-Fault Example

The second species is called argument from negative value.

Premise 1: Value *V* is *negative* as judged by agent *A* (value judgment).

Premise 2: The fact that value *V* is *negative* affects the interpretation, and therefore, the evaluation of goal *G* of agent *A* (if value *V* is *bad*, it goes against commitment to goal *G*).

Conclusion: *V* is a reason for retracting commitment to goal *G*.

How practical reasoning and argument from values are used by the no-fault side in the no-fault insurance example is shown in the Carneades diagram in Figure 7.2. How practical reasoning and argument from values are used by the opposed side in the no-fault insurance example is shown in the argument diagram in Figure 7.3.

Finally, we need to see that one other argument is involved in the deliberations in the no-fault insurance example. One side argues that the no-fault system would have bad consequences by making good drivers pay for bad drivers. The opposed side argues that a no-fault system would fail to lower insurance premiums. Both sides agree that lowering insurance premiums is a good thing, and is even the goal both sides are striving for.

The top argument in Figure 7.3 shows how argument from negative value is used to attack the conclusion that a system of no-fault insurance should be brought in. Below, practical reasoning is used to attack this conclusion. The goal premise of the practical reasoning is supported by a value, illustrating how goals support values in value-based practical reasoning. The other premise, the statement that the new system would lower insurance

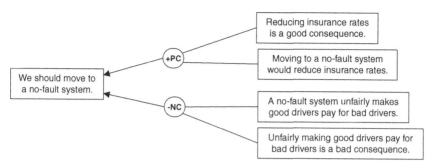

FIGURE 7.4. Argument from Consequences Used by Both Sides in the No-Fault Example

premiums, is attacked by the other side, as shown by the con argument at the bottom left.

To argue that a no-fault system would fail to lower insurance premiums is to argue that such a system would fail to have good consequences. Such an argument is an attack on the practical reasoning of the other side that can be seen as a form of attacking an argument by alleging that it does not have the good consequences it was thought to have. Figure 7.4 shows how argument from consequences is used by both sides in the no-fault insurance deliberation dialogue.

Notice that each side uses argument from consequences to support its proposal. On the right side the party proposing that we should move to a no-fault system uses argument from positive consequences, while on the left side the party opposed to moving to a no-fault system uses argument from negative consequences.

How the local burden of proof shifts depends on the arguments that are used to defend the proposals on each side as the argumentation stage unfolds. The argumentation scheme for practical reasoning may be attacked by critical questions that raise doubts, or by counterarguments. It may also be supported by arguments like argument from values and argument from negative consequences. As the burden of proof dialogue above shows, this local burden of proof shifts back and forth during the dialogue, depending on the moves and countermoves made. However, even though the conclusions of the arguments are actions, or at least statements to the effect that certain states of affairs should be brought about or not, it looks like they are arguments that we can analyze and evaluate within the structure of persuasion dialogue.

5. Analysis of the Wigmore, FDA and Precautionary Principle Examples

We now turn to the second example, Wigmore's example of the man finding a purse in the street (see Chapter 7, Section 1). An explanation of

TABLE 7.2. *Dialogue Structure of Deliberation in Wigmore's Purse Example*

Round	A	M	A's Commitment
1	*A* sees purse. What to do?	*M* claims purse.	Doubt
2	Asks for proof of ownership.	Unconvincing argument.	Doubt
3	*A* hangs on to purse.	*M* describes contents [convincing argument].	Conviction [doubt overcome].
4	*A* hands purse to *M*.	Dialogue concluded.	BoP fulfilled.

how burden of proof operates in the case can be given by seeing the argumentation as a sequence of dialogue. This dialogue structure is shown in Table 7.2.

As the sequence of dialogue represented in Table 7.2 shows, the key factor is how *A*'s commitment to M's argumentation changes during the sequence of moves. When *M* first claims the purse and *A* asks for proof of ownership, *A* is in a state of doubt. Even when an unconvincing argument is presented to him, *A* remains in a state of doubt. It is only at the third move, when *M* describes the contents of the purse, that *A*'s doubt is removed. Thus, the mechanism whereby the burden of proof is fulfilled, and *A* is convinced to hand over the purse to *M*, is the convincing argument presented by *M*. The reason the burden of proof is fulfilled is because the argument presented by *M* meets or supersedes the standard of proof required for *A* to come to accept the proposition that *M* is the owner of the purse. This sequence can be analyzed as an instance of a persuasion dialogue. The proponent claims to own the purse and has the burden of proof. The person who found the purse is the respondent. They are not deliberating about whether or not to give the proponent the purse, but are taking part in a persuasion dialogue about whether or not the proponent owns the purse. This interpretation of the dialogue might still be controversial to some readers, but further discussion of it is provided in Chapter 8.

We now turn to the third example. According to one analysis, burden of proof is obviously very important in the right kind of argumentation that should be required to resolve the issue and solve the problem. According to exponents of this analysis, the example shows that setting the global burden of proof at the opening stage is an important step in solving a problem with deliberation dialogue. This decision would imply that there needs to be a classification of different types of deliberations that distinguishes between ordinary deliberations, like those cited in the previous examples, and special deliberations in cases where public policy decisions need to be made about widely important matters like environmental issues where the anticipated outcome may be at a high level in relation to its impact on public safety and where the decision may have potentially irreversible consequences. These could perhaps be classified as public danger cases. The proposal that appears to be put forward by advocates of the precautionary principle is that public danger cases need to be treated as a special species of deliberation dialogues in which the burden of proof is set to an especially high standard of proof right at the outset, that is, at the opening stage.

According to a second analysis, the deliberation dialogue in this case is actually about the issue of setting the burden of proof in a controversial case. On this analysis, the case is an odd one because the governing question is: on which side should the burden of proof be placed, the FDA or the manufacturers of the pharmaceutical ingredients? Because the dispute itself is about burden of proof, this case does not show that burden of proof in a deliberation type of dialogue should be set at the opening stage. It would appear, according to the advocates of the precautionary principle, the burden of proof to be set in cases of environmental deliberation where the anticipated harm may be a high level should be set at the opening stage. According to this analysis, there is no need to create a special class of deliberation dialogues of the public danger type because if there are serious and widespread consequences that are potentially irreversible, amounting to creating risk of serious public harm, these consequences can be taken into account during the persuasion dialogue in argumentation stage. During this stage, argumentation schemes, like practical reasoning and argumentation from negative consequences, will bring out factors of serious and widespread public consequences that are potentially irreversible. Thus, for example, if there is great danger of irreversible harm to the environment from a particular proposal that has been advocated in environmental deliberations, negative argument from consequences can be used to attack this proposal.

This decision about how to deal with the precautionary principle in terms of formal systems of deliberation by dialogical assignments of burden of proof has implications for the criticism of the principle that it could be applied generally to stop the marketing of any new food products. For those who advocate setting special burdens of proof at the opening stage, the issue becomes one of determining whether a given case, like that of making public decisions on the safety of new food products, can be globally classified as a danger case for not. For those who advocate dealing with the burden of proof locally at the argumentation stage in all cases, the problem is one of weighing safety against matters of which side has greater access to the evidence, matters of setting reasonable standards of proof for safety under conditions of uncertainty and balancing these factors against the value of allowing the introduction of new food products that might have valuable public benefits.

6. The Persuasion Interval in Deliberation

Burden of proof is not the only type of burden one can have in a dialogue. Most of the types of dialogue that have been studied so far in the argumentation literature, such as persuasion dialogue, concern claims that are put forward in the form of a proposition that is held to be true or false. The central aim of the argumentation is to prove that such a proposition is true or false. But other types of dialogue, like deliberation and negotiation, do not have the central aim of proving that a particular proposition is true or false. Still other dialogues are not mainly about argumentation. Some are about the giving and receiving of explanations, for example. In this kind of dialogue, there is no burden of proof because the central aim is not to prove something but to explain something that the questioner claims to fail to understand. However, in this type of dialogue when a questioner asks for an explanation, there is an obligation on the part of the other party to provide one, assuming he is in a position to do that. So generally, in all types of dialogue of the kind that provide normative structures for rational communication, there are obligations to respond in a certain way to a request made in a prior move by the other party. These obligations are quite general, but the notion of burden of proof is more restricted and only applies where a response to an expression of doubt by one party as to whether some proposition is true or not needs to be made by offering an attempt to prove that the proposition is true or false. For obvious reasons, this type of dialogue exchange is centrally important in science and philosophy, but the problem is that the vocabulary used to describe its operation has a tendency to be carried over into other types of dialogue where the central purpose is not to prove or disprove something.

There is no global burden of proof in a deliberation dialogue because no thesis to be proved or disproved is set into place for each side at the opening stage (Walton, 2010). Deliberation is not an adversarial type of dialogue, and at the opening stage all options are left open concerning proposals that might be brought forward to answer the governing question. At the opening stage, the governing question cites a problem that needs to be solved cooperatively by the group conducting the deliberations, a problem that concerns choice of actions by the group. The goal of the dialogue is not to prove or to disprove anything, but to arrive at a decision on the best course of action to take. Hence, the expression "burden of proof" is not generally appropriate for this type of dialogue.

During a later stage, proposals for action are put forward, and what takes place during the argumentation stage is a discussion that examines the arguments both for and against each proposal in order to arrive at a decision on which proposal is best. Something like the standard of proof, called the preponderance of the evidence in law, is operative during this stage. The outcome in a deliberation dialogue should be to select the best proposal, even if that proposal is only marginally better than others that have been offered. A party who offers a proposal is generally advocating it as the best course of action to take, even though in some instances a proposal may merely be put forward hypothetically as something to consider but not necessarily something to adopt as the best course of action. In such instances, it is reasonable to allow one party in a deliberation dialogue to ask another party to justify the proposal that the second party has put forward so that the reasons behind it can be examined and possibly criticized. Hence, there is a place in deliberation dialogue for something comparable to burden of proof. It could be called a burden of defending or justifying a proposal. What needs to be observed is that this burden only comes into play during the argumentation stage where proposals are being put forward, questioned and defended. In contrast with the situation in persuasion dialogue, none of these proposals is formulated and set into place at the opening stage as something that has to be proved or cast into doubt by one of the designated parties in the dialogue. In this regard, persuasion dialogue and deliberation are different in their structures. Because persuasion dialogue (the critical discussion type of dialogue) has been most discussed in the argumentation literature, it seems natural to think that there must be something comparable to burden of proof that is also operative in deliberation dialogue. But this expectation is misleading.

In deliberation dialogue, there is no burden of persuasion set at the opening stage because the proposals will only be formulated as recommendations for particular courses of actions at the later argumentation stage. A deliberation dialogue arises from the need for action, as expressed in a governing question formulated at the opening stage, like "Where shall we go for dinner tonight?" and proposals for action arise only at a later stage in the dialogue (McBurney et al., 2007, 99). There is no burden of proof set for any of the parties in a deliberation at the opening stage. However, at the later argumentation stage, once a proposal has been put forward by a particular party, it will be reasonably assumed by the other participants that this party will be prepared to defend his proposal by using arguments, for example, like the argument that his proposal does not have negative consequences, or the argument that his proposal will fulfill some goal that is taken to be important for the group. How burden of proof figures during the argumentation stage can be seen by examining some of the permissible locutions (speech acts allowed as moves). One of these is the *ask-justify* locution (McBurney et al., 2007, 103).

The locution "ask_ justify" (*Pj, Pi, type, t*) is a request by participant *Pj* of participant *Pi*, seeking justification from *Pi* for the assertion that sentence *t* is a valid instance of type *type*. Following this, *Pi* must either retract the sentence *t* or shift into an embedded persuasion dialogue in which *Pi* seeks to persuade *Pj* that sentence *t* is such a valid instance.

What we see here is that one participant in a deliberation dialogue can ask another participant to justify a proposition that the second party has become committed to through some previous move of a type such as an assertion or proposal. As long as the proposition is in the second party's commitment set, the first party has a right to ask him to justify it or retract it. But notice that when the second party offers such a justification attempt, the dialogue shifts into an embedded persuasion dialogue in which the second party tries to persuade the first party to become committed to this proposition by using a valid argument. So what we see here is that burden of proof is involved during specific groups of moves at the argumentation stage, but when the attempt is made by the respondent to fulfill the request for justification, there is a shift to persuasion dialogue. By this means, the notion of burden of proof appropriate for the persuasion dialogue can be used to evaluate the argument offered.

A key factor that is vitally important for persuasion dialogue is that the participants agree on the issue to be discussed at the opening stage. Each party must have a thesis to be proved. This setting of the issue is vitally important for preventing the discussion from wandering off, or by shifting the burden of proof back and forth and never concluding. In deliberation dialogue however, the proposals are not formulated until a later stage. It makes no sense to attempt to fix the proposals at the opening stage because they need to arise out of the brainstorming discussions that take place after the opening stage. Burden of proof only arises during the argumentation stage in relation to specific kinds of moves made during that stage, and when it does arrive there is a shift of persuasion dialogue, which allows the appropriate notion of burden of proof to be brought in from the persuasion dialogue. Hence, we see that burden of proof plays only a very small role in deliberation dialogue itself. The role it performs is best described not as a burden of proof but as a burden of justification.

7. Conclusions on Burden of Proof in Deliberation

As suggested by the analyses of the examples earlier, burden of proof becomes relevant only during the argumentation stage where there has been a shift from deliberation to a persuasion dialogue interval. It is during this interval, when proposals are put forward and attacked by arguments like argument from negative consequences and argument from negative values, that the need to differentially impose burden of proof becomes operative. For these reasons, it is argued here that no burden of proof should be set at the opening stage of a deliberation dialogue. When competing proposals are brought forward during the argumentation stage, the one to be accepted at the closing stage is the one most strongly supported by the evidence brought forward during the argumentation stage. This criterion corresponds to the proof standard called "best choice" by Gordon and Karacapilidis (1997, 15). A choice is said to meet this standard if no other alternative currently has better arguments. As noted by Atkinson and Bench-Capon (2007, 108), of the five standards of proof set by Gordon and Karacapilidis, the best choice and the "no better alternative" standards apply to deliberation, as contrasted with the other three standards, scintilla of evidence, preponderance of evidence and beyond a reasonable doubt, that apply to persuasion dialogue and matters of fact, as opposed to actions. However, it is argued here that the best choice standard of proof is the one a successful proposal needs to meet during the argumentation stage, except at points where a move in the dialogue indicates that a different standard is appropriate. This standard is that of the standard of the preponderance of the evidence that is used in persuasion dialogue, in that both adopt the standard of a successful proposal (claim) as the one that has more weight supporting it than any other proposal (claim).

In a persuasion dialogue, global burden of proof is defined as a set $\{P, T, S\}$ where P is a participant, T is the thesis to be proved by a participant and S is the standard of proof required to make that proof successful at the closing stage. Burden of proof in a deliberation dialogue defining the standard of proof required to secure victory for a proposal only comes into play during the argumentation stage of the deliberation, once a shift to persuasion dialogue has been made. The standard appropriate for proving it will generally be that of the preponderance of the evidence. To determine whether this standard is met, the argumentation for each of the competing proposals has to be weighed in a comparative manner so that some are judged stronger than others. If there is one that is the strongest, that is the proposal to be accepted, according to the preponderance of the evidence standard set during the argumentation stage.

Support for this approach can be found in a remark of McBurney and Parsons (2001, 420) to the effect that in a deliberation dialogue, the course of action adopted by the participants may only emerge during the course of the dialogue itself, that is, during what is called earlier in this chapter the argumentation stage of the dialogue. It is a corollary of this approach that burden of proof in deliberation dialogue is operative only at the argumentation stage and works in the same way local burden of proof operates in a persuasion dialogue. Once a party has put forward a proposal, he is obliged to defend it, or he can be required to retract it if he is unable to offer a suitable defense.

It is concluded that the burden of proof should not be set more highly against one side than the other in a deliberation dialogue, even in the special type of case where serious harm to the public is at stake. The distinction between such a case and the normal case of a deliberation does not need to be drawn at the opening stage, and can be handled perfectly well during the argumentation stage, as shown by the four examples analyzed earlier. On this model, a deliberation always has the burden of proof set equally during the argumentation stage, so that each side, whatever proposal it puts forward to solve the problem posed by the governing question, has to support its proposal by an argument shown to be stronger than that put forward by the competing side, in order to prove that its proposal is the one that should be accepted. When some evidence of serious irreversible harm to the public is shown to be a possible outcome of a proposal that has been put forward during the argumentation stage, this evidence now becomes a strong argument against the proposal. These factors of serious harm arising as negative consequences of a proposal being considered come out in the argumentation stage, as shown very well in the analyses of the examples presented using argumentation schemes and other tools widely used in persuasion dialogues.

According to this analysis, in such cases, there is a local burden of proof on both sides during the argumentation stage, but the burdens are distributed unequally. The opponent who alleges that there is serious irreversible harm to the public as a consequence of the proposal put forward by the proponent has to use argument from negative consequences. Because the opponent has put forward this argument, in order to make it plausible, he has to fulfill a local burden of proof to give some evidence to support it. At minimum, his argument has to meet the standard of scintilla of evidence to have any worth in shifting the burden of proof to the proponent's side. Once the burden has shifted, the proponent has to give some evidence of safety to a threshold depending on three factors.

- The first factor is how serious the harm is.
- The second factor is how likely the harm is to occur.
- The third factor is what benefits there may be of the positive action that might be weighed against the alleged harm.

For example, to illustrate the third factor, the proposed action may involve the saving of human lives. This kind of argumentation does involve burden of proof because there is a balance between the two sides. When the opponent puts forward even a small bit of evidence there may be serious irreversible harm to the public as a result of implementing the proponent's proposal, the proponent must respond by meeting higher standard in giving an argument for safety based on the three factors cited above. Such matters of burden of proof come into play only during the argumentation stage, once there has been a shift to persuasion dialogue. This kind of argumentation can be represented adequately by the use of the schemes for practical reasoning, argument from values and argument from consequences, as shown in the earlier example.

8. Burden of Proof in Inquiry and Discovery Dialogues

The type of dialogue where use of the expression "burden of proof" is most clearly appropriate is the inquiry. The aim of the inquiry is to collect sufficient evidence to either definitively prove the proposition at issue, or to show that it cannot be proved, despite the exhaustive effort was made to collect all the evidence that was available. The central aim of the inquiry is proof, where this term is taken to imply that a high standard of proof has been met. The negative aim of the inquiry is to avoid later retraction of the proposition that has been proved. And so the very highest standard of proof is appropriate. The inquiry is therefore the model of dialogue in which the expression "burden of proof" has a paradigm status.

The inquiry as a type of dialogue is somewhat similar to the type of reasoning that Aristotle called a demonstration. On his account (*Posterior Analytics*, 1984, 71b26), the premises of a demonstration are themselves indemonstrable, as the grounds of the conclusion, and must be better known than the conclusion and prior to it. He added (*Posterior Analytics*, 1984, 72b25) that circular argumentation is excluded from a demonstration. He argued that because demonstration must be based on premises prior to and better known than the conclusion to be proved, and because the same things cannot simultaneously be both prior and posterior to one another, circular demonstration is not possible, at least in the unqualified sense of the term 'demonstration.'

In contrast, persuasion dialogues, as well as deliberation dialogues and discovery dialogues, have to allow for retractions. It is part of the rationality of argumentation in a persuasion dialogue that if one party proves that the other party has accepted a statement that is demonstrably false, the other party has to immediately retract commitment to that statement. It does not follow that persuasion dialogue has to allow for retractions in all circumstances but, the default position is that it is presumed that retraction should generally be allowed, except in certain situations. In contrast, in the inquiry, the default position is to eliminate the possibility of retraction of commitments, except in certain situations.

Cumulativeness appears to be such a strict model of argumentation that many equate it with the Enlightenment ideal of foundationalism of the kind attacked by Toulmin (1964). To represent any real instance of an inquiry, it is useful to explore inquiry dialogue systems that are not fully cumulative. Black and Hunter (2007) have built a system of argument inquiry dialogues meant to be used in the medical domain to deal with the typical kind of situation in medical knowledge consisting of a database that is incomplete, inconsistent and operates under conditions of uncertainty. This kind of the inquiry dialogue they model is represented by a situation in which many different health care professionals are involved in the care of the patient and who must cooperate by sharing their specialized knowledge in order to provide the best care for the patient. To provide a standard for soundness and completeness of this type of dialogue, Black and Hunter (2007, 2) compare the outcome of one of their actual dialogues with the outcome that would be arrived at by a single agent that has as its beliefs set the union of the beliefs sets of both the agents participating in the dialogue. Their model assumes a form of cumulativeness in which an agent's belief set does not change during a dialogue, but they add that they would like to further explore inquiry dialogues to model the situation in which an agent has a reason for removing a belief from its beliefs it had asserted earlier in the dialogue (Black and Hunter (2007, 6). To model real instances of argumentation inquiry dialogue, it would seem that ways of relaxing the strict requirement of cumulativeness need to be considered.

One difference between burden of proof in inquiry and persuasion dialogues is that the standard of proof generally needs to be set much higher in the inquiry type of dialogue. A similarity between the two types of dialogue is that the burden of proof, including the standard of proof, is set at the opening stage.

Discovery dialogue was first recognized as a distinct type of dialogue different from any of the six basic types of dialogue by McBurney and Parsons (2001). On their account (McBurney and Parsons, 2001, 4), discovery dialogue and inquiry dialogue are distinctively different in a fundamental way. In an inquiry dialogue, the proposition that is to be proved true is designated prior to the course of the argumentation in the dialogue. Whereas in a discovery dialogue, the question of whether truth is to be determined only emerges during the course of the dialogue itself. According to their model of discovery dialogue, participants begin by discussing the purpose of the dialogue, and then during the later stages they use data items, inference mechanisms and consequences to present arguments to each other. Two other tools they use are called criteria and tests. Criteria, like novelty, importance, cost, benefits and so forth are used to compare one data item or consequence with another. The test is a procedure to ascertain the truth or falsity of some proposition, generally undertaken outside the discovery dialogue.

The discovery dialogue moves through ten stages (McBurney and Parsons, 2001, 5) called open dialogue, discuss purpose, share knowledge, discuss mechanisms, infer consequences, discuss criteria, assess consequences, discuss tests, propose conclusions and close dialogue. The names for these stages give the reader some idea of what happens at each stage as the dialogue proceeds by having the participants open the discussion, discuss the purpose of the dialogue, share knowledge by presenting data items to each other, discuss the mechanisms to be used, like the rules of inference, build arguments by inferring consequences from data items, discuss criteria for assessment of consequences presented, assess the consequences in light of the criteria previously presented, discuss the need for undertaking tests of proposed consequences, pose one or more conclusions for possible acceptance and close the dialogue. The stages of the discovery dialogue may be undertaken in any order and may be repeated, according to (McBurney and Parsons, 2001, 6). They add that agreement is not necessary in a discovery dialogue, (2006, 1) unless the participants want to have it.

McBurney and Parsons also present a formal system for discovery dialogue in which its basic components are defined. A wide range of speech acts (permitted locutions) that constitute moves in a discovery dialogue include the following: propose, assert, query, show argument, assess, recommend, accept and retract. There is a commitment store that exists for each participant in the dialogue containing only the propositions that the participant has publicly accepted. All commitments of any participant can be viewed by all participants. They intend their model to be applicable to the problem of identifying risks and opportunities in a situation where knowledge is not shared by multiple agents.

To be able to identify when a dialectical shift from a discovery dialogue to an inquiry dialogue has occurred in a particular case, we first of all have to investigate how the one type of dialogue is different from the other. Most importantly, there are basic differences in how burden of proof, including the standard of proof, operates. In an inquiry dialogue the global burden of proof, that is operative during the whole argumentation stage, is set at the opening stage. In a discovery dialogue no global burden of proof is set at the opening stage that operates over both subsequent stages of the dialogue. McBurney and Parsons (2001, 418) express this difference by writing that in inquiry dialogue, the participants "collaborate to ascertain the truth of some question," while in discovery dialogue, we want to discover something not previously known, and "the question whose truth is to be ascertained may only emerge in the course of the dialogue itself." This difference is highly significant, as it affects how each of the two types of dialogue is fundamentally structured.

In an inquiry dialogue, the global burden of proof is set at the opening stage and is then applied at the closing stage to determine whether the inquiry has been successful or not. This feature is comparable to a persuasion dialogue, where the burden of persuasion is set at the opening stage (Prakken and Sartor, 2007). At the opening stage of the inquiry dialogue, a particular statement has to be specified, so that the object of the inquiry as a whole is to prove or disprove this statement. In a persuasion dialogue, this burden of proof can be imposed on one side, or imposed equally on both sides (Prakken and Sartor, 2006). However, in an inquiry dialogue there can be no asymmetry between the sides. All participants collaborate together to bring forward evidence that can be amassed together to prove or disprove the statement at issue. Discovery dialogue is quite different in this respect. There is no statement set at the beginning in such a manner that the goal of the whole dialogue is to prove or disprove this statement. The basic reason has been made clear by McBurney and Parsons. What is to be discovered is not known at the opening stage of the discovery dialogue. The aim of the discovery dialogue is to try to find something, and until that thing is found, it is not known what is, and hence, it cannot be set as something to be proved or disproved at the opening stage as the goal of the dialogue.

9. Information-Seeking Dialogue, Negotiation and Eristic Dialogue

There seems to be little to say about burden of proof in information-seeking dialogues at first sight, but there are at least two ways in which burden of proof might enter into this type of dialogue. Information dialogue is not exclusively taken up with the putting forward of *ask* and *tell* questions, or with the kind of searching for information one might do when using Google. One reason is that there is a concern not only with obtaining raw information, but with determining the quality of this information by judging its reliability. Judgments of reliability of collected information would seem to involve standards of proof, and therefore also may involve burdens of proof. Another reason is that in many instances of information seeking dialogue, the requesting agent needed to provide the responding agent with an argument in order to obtain access to the information requested. As noted in (Doutre et al., 2006), such dialogues may be viewed as consisting only of *ask* and *tell* locutions if this argument component of them is not considered. But if this argument component is considered part of the information-seeking dialogue, then burden of proof is involved.

There also seems to be little to say about burden of proof in either negotiation dialogue or eristic dialogue, at least that I am aware of, but the reason may be that burden of proof is not an appropriate requirement in either of these types of dialogue. Anyone who adopts the approach to prove something to the other party by means of evidence that fulfills a burden of proof would be likely to perform very badly in either of these types of dialogue. For proving something by using evidence to support your claims should not be the central goal in either of these types of dialogue. However, in both types of dialogue there are typically intervals where there is a shift from one to another type of dialogue where burden of proof is important. For example, a contractor may be negotiating a price for installing a new basement in the house, and at some point in the dialogue it may become important for the contractor to try to convince the homeowner that the building code for walls in basements in that area specifies certain requirements that have to be met, for example, ruling on the thickness of the walls. In such a case, the notion of burden of proof may not play any direct role in the negotiation argumentation itself, but when there is a shift from it to a persuasion dialogue where the contractor tries to convince the homeowner that walls of a certain minimum thickness are mandatory, burden of proof may be an important factor in evaluating his arguments. It may be, as well, that when agents argue about receiving permission to get information during an information-seeking dialogue, there has been a shift to some other type of dialogue, for example a persuasion dialogue.

10. The Contextual Nature of Burden of Proof

One of the problems with the notion of burden of proof is that it is highly confusing. Althought it is clear enough in some cases, it is often applied to other cases in a way that makes sense only by transference. It has been shown in this book that burden of proof is contextual – it depends on the type of dialogue an argument is supposedly part of. This contextual variation is an important reason why the notion of burden of proof rightly is thought to be so ambiguous and slippery.

Burden of proof is vitally important in the inquiry type of dialogue. The goal of the inquiry is to collect a sufficient quantity of evidence to prove a designated proposition to a high enough standard of proof so that it will not need to be retracted later. This proposition is designated at the opening stage. All the argumentation in the inquiry moves forward in a cumulative fashion so that there will be a buildup of evidence. The inquiry aims to provide proof of a particular proposition designated at the opening stage. Alternatively, the inquiry can also fulfill its goal if it can prove that this designated proposition cannot be proved. The inquiry can carry out this aim if it can marshal enough solidly established evidence to show, to a high standard of proof, that this proposition cannot be proved. The inquiry can be closed either way.

The term "burden of proof" is misleading in persuasion dialogue because the outcome of a persuasion dialogue is never a conclusive proof of a proposition. What a persuasion dialogue does is to evaluate the evidence on both sides of a disputed issue to arrive at a conclusion that shows which side has the strongest argument. The aim of a persuasion dialogue is to use the commitments of the other party or the audience to accept the view that you advocate and that they are opposed to or question. Hence, the term "burden of persuasion" is particularly appropriate for use in this context.

The term "burden of proof" is also inappropriate in deliberation dialogue, because deliberation dialogue isn't about proving that some proposition is true or false. It is about choosing the best course of action from a number of alternatives. Proposals are put forward, and after the pros and cons of each proposal has been brought up and discussed, the decision is arrived at on the issue of which proposal is the best to accept. The rational way to proceed in a deliberation is to select the best proposal as the means to move ahead with a plan or course of action. A proposal is not proved to be true or false, and so there is no burden of proof for or against it. However, just as there is said to be a burden of proof to defend a claim with evidence when it is questioned in a persuasion dialogue, there is a reasonable expectation to defend a proposal that one has put forward by giving reasons to support the claim that it is a good proposal. However, just as in persuasion dialogue, it is not clear that this expectation is highly significant, because each party is trying to put forward a proposal that will be accepted.

This expectation is comparable to what is called the burden of producing evidence in a persuasion dialogue.

As an example to support their contention that burden of proof is operative in deliberation contexts (2007, 43), Hahn and Oakesford (2007) cite the decision that many countries have had to face when deciding whether or not to sign up for the Kyoto Agreement. The majority of papers in leading scientific journals have accepted the claim that global warming is real, even though debate on the topic continues. However, they write (43), "the possible consequences of global warming are so potentially devastating that one might not want to wait until one was entirely certain before taking action." Accordingly, the procedure governments use is to set a threshold for action so that they can arrive at a decision when they are convinced enough to act. This example provides a paradigm case of the use of burden of proof as a device for rational decision making leading to a course of action even under conditions of uncertainty.

The problem with using this example to support their theory is that they use the phrase "convinced enough to act," which suggests a persuasion dialogue over actions. On the other hand, the example is taken by them to represent a deliberation, because the countries are deciding whether or not to sign the Kyoto Agreement, which is a decision between two courses of action that they are faced with. If deliberation is about decision making between courses of action, then burden of proof finds its paradigm use in decision making linked to actions, and this example should be an excellent case to illustrate the use of burden of proof. But there is a significant ambiguity in the example itself. Is the burden of proof notion operative in the persuasion dialogue in which the parties use a threshold to become convinced enough to act, or is it also operative in the deliberation dialogue in which they choose whether to sign the agreement are not? Or is it operative in both? In Chapter 7 it was argued that there is no burden of proof in a deliberation dialogue, and that burden of proof only becomes operative when the deliberation shifts to a persuasion dialogue interval where evidence is used to support factual claims.

One analysis of the example would be the following. The participants who have to decide whether or not to sign the agreement do it on the basis of factual conflict of opinions on whether global warming is real or not. The assumption they work with is to base their decision of whether or not to sign the agreement on the issue of whether global warming is real or not. If it is real they will sign the agreement, but if they are convinced by the evidence of scientific opinions expressed by papers in leading journals that it is not real, they will not sign the agreement. A very nice model of the argumentation in the example is to see it as primarily a persuasion dialogue in which a conflict of opinion needs to be resolved. There is evidence that it is indeed a persuasion dialogue concerning the conflict of opinions on whether global warming is real or not. The evidence on which both sides base their arguments takes the form of argument from expert opinion. The important thing is that there is a conflict of scientific opinions on the issue of global warming, even though presently the majority of scientists are for the global warming hypothesis. If we analyze the argumentation in the example this way, there is a shift from the persuasion dialogue about global warming to a deliberation dialogue on the question of whether to sign the agreement were not. Factors of burden of proof could be significant in both dialogues, but the way they present the example mainly it seems to be operative in the persuasion dialogue, for once the decision is made about which scientific opinion to accept, the decision making in the deliberation dialogue is relatively trivial. However, the way the example is presented, there is some evidence of factors that look like burden of proof operating in the deliberation part of the dialogue. Hahn and Oaksford (43) write that although the thresholds for decision making are determined by many factors in a case like this, some of the factors are the perceived costs and benefits of action and inaction. This part of the description of the example suggests the use of argumentation from consequences, a typical form of argumentation in deliberation dialogue, even though it can also be used in persuasion dialogue. Thus, the example is not without its complications, but the bottom line is that persuasion dialogue is such an important part in it that burden of proof seems mainly to be operative in that part of it, contrary to the theory that the example was used to illustrate.

In a persuasion dialogue, the burden of persuasion operates at a global level, and one might ask whether there is some comparable notion in deliberation that is fixed in place until the closing stage. The answer is that there is a standard of proof set in place at the opening stage of every deliberation dialogue, and that standard is the preponderance of evidence standard. In other words, when all the various proposals for action are put forward during the argumentation stage of a deliberation dialogue, the participants weigh the merits of each proposal comparatively and try to arrive at a reasoned conclusion on which proposal has the strongest reasons in its favor. If there is a tie between two proposals, the rational conclusion is to pick one or the other at random, or for some other reason than the weight of the evidence supporting or detracting from the proposals. So really there is no counterpart to the burden of persuasion that is set at the opening stage of the persuasion dialogue. Or at least the only counterpart is the requirement that the dialogue is to be concluded when one among the various proposals put forward and argued for has been shown to be supported by a greater weight of reasons behind it than all the other proposals. To put it another way, the standard of preponderance of the evidence is set on all parties equally at the opening stage of a deliberation dialogue.

In information-seeking dialogue, the goal is to collect information on a particular topic, where this topic is clearly defined at the opening stage. But it is important that the information that is collected should be reliable, and should not turn out to be misinformation, so to speak. Standards come into play in deciding how reliable the information collected and used needs to be. So although something comparable to burden of proof comes into play, generally the aim of an information seeking dialogue is not to prove or disprove some designated proposition. Therefore, it is not entirely appropriate to speak of a burden of proof in this context. It would be better to speak of a burden of reliability, or a burden of showing that your information is reliable and can be trusted. It could be called a burden of verification of information, or perhaps a burden of accuracy.

In a quarrel, participants will often try to accuse each other of failing to satisfy burden of proof. Or they will try to make a burden of proof appear to shift to the other side, for example, making the other side appear to be guilty of something. But really there is no notion comparable to burden of proof in a quarrel. Burden of proof only comes into eristic dialogue when a participant pretends to be engaging in a higher type of dialogue, such as the inquiry or persuasion dialogue, both types of dialogue where burden of proof is important.

8

Conclusions

Previous chapters have shown that formal dialogue models of argumentation, along with tools from AI like the Carneades Argumentation System and the abstract argumentation framework, apply very well to modeling burden of proof and presumption in the well-organized, rule-directed framework of a legal trial. The objection posed in Chapter 1 was that the legal concepts of burden of proof and presumption have been illicitly transferred from the legal setting to public policy discussions and other arenas where the argumentation is not structured in the same way it is in a legal setting. Chapter 8 takes up this challenge by arguing that the formal dialectical framework of Chapter 4 can be usefully applied to modeling burden of proof and presumption in these other settings. This argument, of course, does not claim the burden of proof and presumption work in every respect in the same way they work in other settings. It only means that the methods used to reason about evidence in the common law system presents an outline of reasonable but defeasible argumentation that has some important features, represented in the dialogue models of burden of proof and presumptions presented in the previous chapters of this book, and these features can be adapted to and applied in a helpful manner to other settings of argumentation outside law.

Of course there are many different settings in which argumentation is used as a means of proving a hypothesis, settling a conflict of opinions based on evidence brought forward or arriving at a rational decision on how to make a choice in a deliberation on what to do in a situation requiring such a choice, as indicated in Chapter 7. For the purposes of this book, however, there are two main settings that seem to be of main interest in relation to the issue of transferability of the legal notions of presumption and burden of proof. One is the kind of organized disputation of the kind represented in a forensic debate, or to cite a specific example, a presidential debate that has a moderator and is televised to a large audience. The other is the setting of ordinary conversational argumentation, which in the paradigm case has only two participants in a discussion on a conflict of opinions concerning some issue such as privacy on the Internet. In this chapter, we will discuss the transferability of the model of burden of proof and presumption built in the previous chapters to these arenas of argumentation.

How do we tell generally in a given case which side has the burden of proof? How this is to be done depends on how much information there is in a given case and what kind of discussion is supposed to be taking place. The normal default rule is that the party who makes a claim has the burden of proof to support that claim if the other party questions it. Thus, in many cases it might be a very simple matter to analyze an example with respect to burden of proof if the example is one where a claim has been made, that is some proposition has been put forward as an assertion, and those of us who are attempting to analyze or evaluate the argumentation in the example question the claim. Other cases, as we have seen in Chapter 2, are not so simple. There may be a lengthy chain of argument presented on both sides, for example in a case of argumentation put forward in legal trial or a forensic debate. In this kind of case we have to confront the problem of how the burden shifts back and forth during the entire sequence of argumentation as the parties on both sides take turns putting forward challenges and counterarguments. However, we have already examined some examples of this sort in legal argumentation in Chapter 2. Now, in Chapter 8, we need to confront the problem of whether this kind of approach can be applied to a case outside law where there is an absence of the same kinds of procedural rules and constraints. To confront this task, we use a fairly simple example that represents a familiar kind of persuasion dialogue in everyday conversation argumentation.

1. The Allegation of Hasty Transference

Hahn and Oaksford (2007, 39) argue that burden of proof is important in law, where the goal is to make practical decisions but though it has been given a central role in normative accounts of argumentation, it has no place of importance in a critical discussion. According to their characterization, the goal of a critical discussion is to increase or decrease degree of belief in a proposition, and in such cases it is not necessarily important that degree of belief reaches a certain threshold representing its standard of proof. For these reasons, they support the view of Gaskins that transferring the notion of burden of proof from a legal setting to a setting of everyday conversational argumentation is inherently flawed as a research program for studying the nature of burden of proof in arguments.

In all three versions of their set of rules for the critical discussion, van Eemeren and Grootendorst set down a particular rule that governs burden of proof. In the 1992 version (van Eemeren and Grootendorst, 1992, 208), the rule governing burden of proof is simple. It only requires that "a party that advances the standpoint is obliged to defend it if the other party asks him to do so." For example, Rule 8a of the formal dialogue system PPD (Walton and Krabbe, 1995, 136) says, "If one party challenges some assertion of the other party, the second party is to present, in the next move, at least one argument for that assertion." Hahn and Oaksford (2007, 47) have questioned whether van Eemeren and Grootendorst need to have Rule 3 requiring burden of proof in a critical discussion. They think it makes sense to have a burden of proof for a participant's ultimate thesis set forth at the opening stage of the critical discussion, but they question why it is useful for each individual claim in the argumentative exchange to have an associated burden of proof. They concede that although there is a risk of nonpersuasion in not responding to a challenge by putting forward an argument to defend one's claim, this risk is a relatively small factor in the outcome of the dialogue and "is entirely external to the dialogue and not a burden of proof in any conventional sense" (Hahn and Oaksford, 2007, 47). They have a point. It is worth asking what function the requirement of burden of proof has in a persuasion dialogue of the type represented by the critical discussion.

The addition of a third party audience to the persuasion dialogue brings out the utility of burden of proof. If a party in a persuasion dialogue puts forward an argument, and then fails to defend it when challenged to do so, this failure will make his side appear weak to the audience who is evaluating the argumentation on both sides. They will ask why he put forward this particular claim if he can't defend it, and he may easily lose by default. This can come about because the audience has the role of being a neutral third party in the dialogue, and is not merely one of the contestants trying to get the best of the opposed party. It helps the audience to judge which side had the better argument if each side responds to challenges by putting forward arguments to support its claims (Bench-Capon and Sartor, 2003). Law is an area where there is such a third party trier (a judge or jury) in addition to the opposed advocates on each side.

Legal argumentation of the kind used in the trial is best modeled as a type of persuasion dialogue with three participants, the advocates for the two opposed sides and a third-party trier (Bench-Capon et al., 2007). The third-party trier, it is shown, falls under the category of an audience, defined as a participant with a carefully defined rule in a three-party persuasion dialogue. The aim of the advocate on each side is to persuade the audience to accept a particular proposition or to cast doubt on the opponent's proposition to be proved. This task is essentially called the "burden of persuasion." It is argued that what makes legal argumentation in the trial distinctive as a type of persuasion dialogue is that it uses the notion of burden of proof to try to ensure, insofar as this is possible, that the dialogue will conclude by enabling the third party to arrive at a reasoned decision for the one side or the other on the basis of the admissible evidence.

Hahn and Oaksford's criticism has a point because the way the critical discussion is currently portrayed in the argumentation literature is a simplified model, one that needs to be expanded in several directions to account for the complexities of legal argumentation. However, it can be extended in a way that clearly brings out the role and importance of burden of proof by adding several new features and distinguishing among different types of persuasion dialogues, one of which is the critical discussion. The basic insight here is that there can be different formal models of persuasion dialogue that can be used for different purposes in studying real cases of argumentation (Krabbe, 2013). The research program undertaken in this chapter is to start with a very simple abstract model of persuasion dialogue, one that reveals its simplest and most basic features, but that needs to be extended and enriched in order to be applied most usefully to both instances of everyday conversational argumentation and legal argumentation. By this means, it is shown, in a much better way than has been possible in the past, how legal argumentation and everyday conversational argumentation are related to each other. Centrally, both involve persuasion dialogue. Indeed, because in many cases, the arguments presented by both sides in a trial are directed toward a third-party audience that is a jury, people who are normally not legal specialists in evidence law, the arguments used by both sides are in fact everyday conversational arguments, even though they are restricted by conditions of admissibility and other procedural requirements of law. What is shown, however, is that although legal arguments in a trial are managed and evaluated in quite a different way from those in everyday conversational arguments, the two share a common structure. The problem is that this structure is much more visible in legal argumentation, where the rules have been made explicit, whereas they are more implicit, and more on the margins, in everyday conversational argumentation. The problem faced here is to see how to better get a grasp of everyday conversational argument, given that its conversational setting is typically less than readily apparent. It was this problem Grice first graphically brought to our attention by showing how conversational rules are implicit in certain kinds of moves we make in everyday conversational argumentation. Advancing beyond that point through the subsequent argumentation literature has taken us further toward solving that problem, but as this book will show, we need to take further steps in order to deal effectively with problems associated with burden of proof and presumption.

As shown in Chapter 1, Hahn and Oaksford (2007, 40) argued that there has been a hasty transference of the notion of burden of proof from its proper domain of application in law. As indicated in Chapter 1, the two premises of their analysis are the propositions that burden of proof is only important where action is concerned, and that questions of evidence in law are subsidiary to decisions about actions. Hahn and Oaksford (2007, 48) also claimed that "termination does not seem essential to argumentative dialogue in general." On these considerations, they draw the conclusion (49) that there is no need for burden of proof in a critical discussion of the kind found outside law in everyday conversational argumentation because this is not a type of dialogue with an inherent link to action.

2. Comparing Legal and Nonlegal Burden of Proof

Here we return to Wigmore's property investment example described in Chapter 1 (Section 3). As noted in Chapter 1 (Section 3), this case is not one that is being argued in a trial setting. It is an argument that takes place in a setting of everyday conversational argumentation, where there is no third-party trier who decides the outcome. Hence, although Wigmore describes it as a case of risk of nonpersuasion, it is different from the kind of case in court where the judge closes off the trial because one party has failed to produce enough evidence to have any chance of defeating the other side. Situating the case with respect to the classification of burdens of proof of Prakken and Sartor, it would be classified as a tactical burden of proof. Nevertheless, perhaps Wigmore is right to see the case by classifying it under the heading of an evidential burden, or as he calls it, one that involves a risk of nonpersuasion type of burden of proof.

The example is also interesting in regard to another fundamental point. Wigmore describes the goals of the two opposed parties in the argumentation in the example as that of persuading the third party of their contentions. He writes that "their desire is respectively to persuade M as to their contention" (285), and he equates the burden of proof with the risk of nonpersuasion. Hence, it would seem reasonable to assume that the argumentation in the example fits the category of a persuasion dialogue. On the other hand, Wigmore also writes (286): "it is the desire to have action taken that is important." In the example, the penalty for A of not fulfilling the burden of proof by not persuading M beyond the doubting point is that M will not take the desired action. Clearly action is involved in an important way in the example, and therefore one might dispute whether the argumentation in the example really fits the structure of a persuasion dialogue. One might claim that it is a deliberation type of dialogue. For deliberation is all about choice between actions that an agent can take in a particular set of circumstances. The question of whether the argumentation in this fundamental kind of case fits the structure of a persuasion dialogue or that of a deliberation dialogue is a highly controversial issue.

The example is indeed a very interesting one to throw light on the issue of what role burden of proof plays in persuasion dialogue and deliberation dialogue generally. We can look at the dialogue as a contest of opposed goals in different ways. In one way we can look at it as about actions. A is trying to get M to invest money in his property, and B is opposed to having

M take this action. In another way we could describe the structure of the dispute by saying that *A* is trying to persuade *M* that *B*'s investing in *A*'s property is a good idea. Wigmore's wording of the example where he writes "*M* will invest in *A*'s property if he can learn that it is a profitable object and not otherwise" suggests that the issue is about whether the proposed investment will be profitable or not. This way of describing the argumentation in the example suggests a persuasion dialogue about actions. *A*'s ultimate conclusion to be proved in the dialogue is that investing in his property is a good idea, based on the implicit assumption that something is a good idea if it is profitable. This way of describing the argumentation has it that the proposition that the proposed investment will be profitable is being offered as a reason to persuade *M* to accept the proposition that investing is a good idea. In this respect, the property investment example can be compared to Wigmore's purse example described in Chapter 7, Section 1.

Wigmore (1981, 286) did pose the question of what the differences are between burden of proof in litigation and burden of proof "in affairs at large" outside the legal setting. His answer was that the procedures and penalties are different in litigation, but these differences are minor compared to what he called a single "radical difference." He called this difference (286) "the mode of determining the propositions of persuasion which are a prerequisite" to the actions of the third-party trier (audience). What did he mean by this? Basically he meant that there are laws of pleading and procedure that assign tasks and obligations to one or the other party as prerequisites for getting a favorable outcome from the trier. For example, the law defines what needs to be proved (the elements) in order for the prosecution to win in a murder trial, usually killing and guilty intent. The law also specifies what needs to be shown by the defense in order to persuade the tribunal to reverse its action, that is, the law specifies exceptions that constitute an excuse or justification. In other words, on Wigmore's view, burden of proof works basically the same way in law as in arguments on practical affairs outside of law, except that law narrows the groups of propositions that need to be proved for one side to obtain a favorable ruling of the trier, and kinds of arguments that the other side can use to reverse a favorable ruling.

In rebuttal of Gaskins' views about burden of proof and presumption in law Allen (1994) showed in detail how American evidence scholars have studied these concepts in depth and have built a body of knowledge about them that does provide clear standards on how they should be used in legal argumentation. Allen replied (1994, 629) that the Gaskins negative description of how burden of proof works in law as a shadowy device used by skillful advocates in legal battles to direct arguments from ignorance against each other is not accurate as a way to characterize prevalent practices of argumentation in law. Allen replied (630) that the rules of law that virtually always preexist in a trial fix the burden of proof, the so-called burden of persuasion, during the opening stage. Thus, even though burden of proof does shift back and forth during the argumentation stage, this shifting takes place in a stabilized manner that follows a controlled logical procedure. Allen (632) shows that burden of proof and presumption are useful tools that the legal system uses to deal with the problem of lack of evidence concerning disputed events. In the common law system, the parties to the dispute themselves are responsible for providing the information. What is needed is some "mechanism for structuring the orderly presentation of that information so that a decision can be reached" (633). The tool used for this purpose in law is the burden of production, referring to the obligation of a party to bring forward enough evidence to show that a legitimate dispute exists and thereby keep a trial moving forward. How strong should such evidence need to be? Allen replies that it needs to be "at least strong enough so that reasonable people could disagree about who should win the case" (633). Whether reasonable people could disagree about who should win the case is determined by the burden of persuasion, set at the opening stage of the trial.

On Allen's analysis of burden of proof, the defining trait of litigation is the problem of arriving at a decision under conditions of uncertain knowledge (633). You can see straight away from this assumption that argument from ignorance is not only a legitimate form of argument in legal argumentation, its employment is a necessary tool in that setting. Allen's explanation of the structure of the argumentation is revealing, in that it stresses that the decision concerning the outcome of the trial is reached by a third-party audience that evaluates reports of events brought forward into the trial setting by other parties. These reports might not only be in error, but also they often tend to be opposed to each other, given that the rationale of the trial is to resolve a dispute between two sides. Decision making under uncertainty for these reasons, is carried out by setting standards of proof, such as preponderance of the evidence, that are not shadowy, as suggested by Gaskins, but are precise. These standards are well articulated and made known in advance to all participants in a trial. To see how burden of proof works in a trial in a precise way to moderate the argumentation on both sides and that is fair to the litigants in allowing the evidence for and against their claims to be presented and evaluated, the most important factor is the relationship between the two kinds of burden of proof, burden of persuasion and burden of production. What is important is to see that the two types of burden are linked together. As Allen puts it (634), the burden of production needs to be seen as a function of burden of persuasion: "A burden of production is satisfied when a reasonable person applying the relevant burden of persuasion could find in favor of the person bearing the burden of persuasion." In other words, once the burden of persuasion is set in a given case, it is the means whereby the burden of production can then be evaluated as the argumentation proceeds through the trial. Although the burden of proof shifts back and forth from one side to the other during the argumentation stage, this process takes place in a controlled fashion. Not only that, but when problems arise concerning which side the burden should be placed on, the judge can intervene, and make a ruling.

Allen's careful and well-documented analysis of how burden of proof and presumption work in legal argumentation rebuts the basis of many of Gaskins' arguments to the effect that burden of proof is a shadowy concept that is utilized by lawyers and judges in legal argumentation in a suspicious, uncontrolled shadowy way to manipulate the decision-making process to suit some hidden agenda. It offers a clear, succinct and accurate summary of the essentials of how burden of proof really works in legal argumentation as a tool for arriving at reasoned decisions under conditions of uncertainty, conflicts of opinion and lack of complete information. It also provides good evidence to suggest that the use of argument from ignorance, presumption and burden of proof are the kinds of tools needed to deal with argumentation in the most common kinds of situations where there is a conflict of opinions because there is a lack of evidence of the kind that would arrive at a decision directly on what conclusion to draw. This approach offers some hope and direction for those of us attempting to provide logical models of these forms of argumentation that could be used to help provide a sound basis for determining in real cases of argumentation when a given argument is reasonable or not, and how strong it should be taken to be within procedural setting that can be objectively laid out.

3. Normative Models and Everyday Conversational Arguments

It is an interesting question to ask what the relationship is between an abstract normative model of dialogue, and an everyday conversational argument that may be highly unstructured. In everyday conversational argumentation, arguments are often put forward in a situation where there has been no agreement beforehand on what the issue is that the discussion is supposed to be about. We can have the following sort of exchange, for example.

Bob: I think it's going to rain today.
Alice: What makes you think that?
Bob: Well, you can see that there are dark clouds in the sky.

At his first move, Bob ventures an opinion. Alice then asks him for a reason to support his opinion. Bob follows by giving a reason. This dialogue may have been part of a larger dialogue, a discussion about the weather or some other topic. But it may not be. They may have been talking about something else, not talking about anything related to the weather, and out of the blue, so to speak, Bob ventured this opinion. Should we say that by venturing the opinion that it's going to rain today, Bob has started a new persuasion dialogue, setting in place the issue of whether it's going to rain today or not? Or should we say that there really is no issue at stake. Perhaps, for example, the conversation about the weather may end at that point. In effect, there is no opening stage or closing stage, and whether there is or is not doesn't really matter. Bob responded to Alice's request to give a reason to support his opinion, and there may be no need to try to determine how strong Bob's argument is, or whether Alice should accept it.

We might conclude from this example, and many others like it, that generally in everyday conversational argumentation, unlike a forensic debate or legal trial, there is no issue set at the beginning of the discussion, along with a pro and contra determination to be set for each side. If we take this claim to be true, we might draw an inference from it to the conclusion that burden of proof has no significant role in everyday conversational argumentation. This inference is, however, dubious, along with the conclusion that supposedly follows from it. What kind of evidence is there that people, when they are engaged in everyday conversational argumentation, accept or rely on some kind of overarching structure that can be applied to their conversational moves? There is really no empirical evidence of this sort that can be sought, at least directly, but there are four kinds of indirect evidence that can be drawn by inference from the way people conduct themselves in ordinary conversational argumentation.

The first source of evidence can be found in the rhetorical maxim that framing the issue of the discussion is extremely important in everyday conversational argumentation, for example, of the kind we commonly encounter in political discussions. During the course of an argument, one of the participants will sometimes attempt to reframe the issue by formulating the issue of the discussion in a way different from the way it appears to have been formulated. This move is often said to be a powerful rhetorical tactic. One reason it is so powerful is that it changes which arguments can be considered to be relevant or not.

There are also three kinds of indirect evidence to be found when participants feel that the other party has somehow made a wrong or unfair move that ought to be objected to as procedurally inappropriate. There are many kinds of common objections of the sort that could be cited. The first of the three cited here is the objection of irrelevance made by one party who objects that the other party has strayed away from the issue that the discussion was originally supposed to be about. This kind of objection implies that the conversation had a topic or issue that was identified at some previous point, or that was implied by some move that opened the discussion leading to the present point. For example, where there has been a shift from one topic to another during the same sequence of argumentation, one party might object to the new line of argumentation as not relevant to the topic that is supposedly being discussed. People commonly make this kind of objection by saying things like, "We are supposed to be talking about topic X, but now you have shifted to topic Y." The implication of this objection is that the discussion about topic X will never be

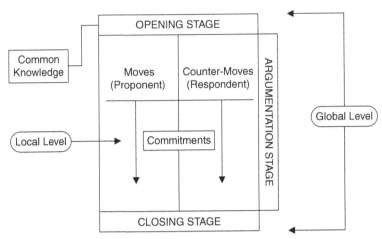

FIGURE 8.1. General Structure of a Dialogue Procedure

resolved if the argumentation shifts to a different topic Y and therefore now fails to bear the probative weight concerning the proposition to be proved or refuted on topic X. The implication is that if the argumentation doesn't continue along the lines toward resolving the issue of topic X, the discussion about topic X will never be appropriately concluded. This kind of objection is related to arguments about reframing an issue. It has to do with the issue to be discussed that needs to be stayed with, once an argument about it has started.

A framework for rational argumentation, whether it is of the persuasion type of dialogue or not, has a certain minimal normative structure. This structure allows it to make sense as an orderly procedure. There is an opening stage. Then the dialogue moves to an argumentation stage where the parties take turns to make their moves. At a closing stage the dialogue is terminated. This general structure is shown in Figure 8.1. Everyday conversational argumentation sometimes wanders around and seems to have no direction. An argument on one topic starts out and then at some point shifts to an argument on another topic, or even to another type of discourse different from argumentation, such as offering an explanation. It seems that conversational argumentation has no structure, like a beginning point or an ending point, so that it makes sense as a connected sequence with a direction.

How burden of proof fits into the normative structure of an orderly dialogue procedure was already formulated in a general way after the pattern outlined in Figure 8.1 (Walton, 1988, 240). A reasoned dialogue was defined as an ordered sequence of pairs of moves that begins at an initial move or opening stage, and proceeds toward a final move, or closing stage. This view is a modified version of the insight of van Eemeren and Grootendorst (1984) who distinguished an initial phase of the critical discussion they called the confrontation stage, where the participants articulate the basic conflict of opinions that is the basis of the dialogue and clarify or agree on some of the procedural rules. Agreements set in place at the opening stage define the purpose and scope of the dialogue in a global manner. The global agreements set at the opening stage stay in place and govern the whole dialogue. Local moves made during the argumentation stage influence how the burden shifts back and forth between the protagonist and the antagonist as arguments are put forward by one side and challenged by the other side (Krabbe, 2013). The thread of continuity in the local argumentation sequence is that each side must respond to the previous move made by the other. There has to be continuity, and the sequence of argumentation must be aimed at resolving the original conflict.

The fact that people pay attention to the kind of criticism based on allegations of irrelevance, and see it as a legitimate objection, is another kind of evidence that sticking to a topic long enough to consider the arguments on both sides deeply enough to come to some kind of conclusion about that topic is seen as a proper requirement of rational argumentation. A second kind of objection is made when one party feels the other is monopolizing the conversation, or even worse, not letting him speak. This objection could be put in the form, "Let me speak now." It appeals to the normative ideal that argumentation in a dialogue format should have the characteristics of turn taking. This ideal is that once one party has made a move, the other party should be allowed to respond to that move. This ideal is expressed in the pragma-dialectical rule to the effect that neither party should prevent the other party from speaking. Yet another kind of objection is "There is no evidence for what you said." This objection is what we call a challenge, and it expresses the idea of burden of proof for an assertion.

People do see irrelevant argumentation as problematic, and even associated with logical fallacies of relevance. The existence of the tradition of informal fallacies generally provides evidence that even in everyday conversational argumentation of the kind one finds in newspapers and magazines for example, which often seems fragmented and unstructured, it is implied that generally people expect to see certain procedural and logical rules followed. Of course, that expectation is not always met, and invoking the need for it often does not even arise in everyday conversational argumentation.

Hence, although it may seem that ordinary conversational argumentation is very often chaotic and disorganized, and therefore admits of no rules, this initial appearance is undercut by the recognition that participants do often make objections to the effect that the conversation is not following appropriate procedural rules for rational argumentation. The bottom line is that people do see argumentation as having a purpose, and they do see it as having to conform to procedural rules in order to achieve this purpose.

In some cases, it is clear that the participants are just quarreling, and there is no real expectation of following rules requiring rational argumentation, except perhaps rules of the most minimal sort, like taking turns. In most instances, however, a participant or an observer can or will object to some kind of move being made that violates requirements for rational argumentation. People expect that if an argument has started about some particular issue that has been identified as worthy of discussion, and the respondent does not immediately accept the argument that has been put forward by the proponent, the argumentation will continue long enough at least until some pro and contra reasons have been given. Otherwise, the argument has not been given a fair hearing and consideration, and is simply left hanging. While it may not be possible to settle the argument, at least we feel there has been some value to the discussion if the reasons that can be brought forward for and against it can be expressed. At least then we are in a position to make up our own minds what to think.

The presence of such rules of procedure is visible in forensic debate or in legal argumentation, where there are penalties for violating them. In the case of legal argumentation in a common law trial setting, for example, an argument can be ruled as inadmissible if it carries no probative weight with respect to the ultimate thesis to be proved by its proponent. Because the rules of argumentation in a trial are stricter, it is easier to make a case that no comparable rules apply to everyday conversational argumentation, such as the rule requiring relevance. However, the presence of an implied dialogue structure governing everyday conversational arguments is indirectly implied by the four kinds of evidence cited earlier, and other comparable evidence. These kinds of evidence relate to informal fallacies. For example, if one party tries to criticize the arguments of another by imputing premises to him that do not fairly represent his commitments, the first party can and should object. This kind of objection is related to the straw man fallacy. It is implied that commitments of both parties are being recorded and managed by keeping a log of what each party previously said. Of course, this expectation is only a normative assumption, and the problem in many cases is that nobody keeps track of what was said, and an argument often arises on that very issue. Still, the legitimacy of such an objection implies the existence of a normative structure like that represented in Figure 8.1.

4. The Dialogue on Tipping

The following example represents a shorter sequence of exchanges adapted from a longer dialogue on the issue of tipping (Walton, 1992, 7–8). During a dinner party, the host related a story about his experience when he was worried about how much money he was expected to give as a tip to the bellhop in a hotel. Two participants in the dinner party, Helen and Bob, begin to discuss the matter, and it became apparent that they disagreed with each other on the subject of whether tipping is a good practice or not. They agreed to dispute the issue, and took turns putting forward arguments on each side. This dialogue can serve as a manageable example to illustrate how the technology applied to legal reasoning in Chapter 2 can be adapted to also apply to modeling how burden of proof shifts from side to side in a persuasion dialogue in everyday conversation argumentation.

The issue of the dialogue is whether tipping is a good or a bad practice. Bob's arguments support the claim that tipping is a good practice while Helen argues for the opposite claim. We will assume that the proposition 'tipping is a good practice' will be interpreted in such a way that it is the negation or opposite of the proposition "tipping is a bad practice." We will also assume that the standard of proof for the dialogue is that of preponderance of the evidence. This makes sense if we assume that once the arguments on each side are aggregated together into a long sequence of argumentation, the stronger argument is taken to prevail. In other words, to see which side wins at the end of the discussion, we will have to examine the lengthy sequence of argumentation on each side, one consisting of Bob's arguments to support his thesis, and the other consisting of Helen's arguments to support her thesis, and make a determination of which side has put forward the stronger line of argumentation.

Performing this task for the lengthy sequence of argumentation in the original dialogue on tipping in (Walton, 1992, 7–8) would be a lengthy case study. Luckily however, it is not necessary to analyze and evaluate such a lengthy and complex sequence of argumentation as that contained in the original example in order to give the reader a clear idea of how this task can generally be done. What we do here is take a shorter and simplified part of the argumentation consisting of the first few dialogue exchanges, and show how it can be modeled using the Carneades Argumentation System.

We assume at the beginning of the discussion that both Bob and Helen have agreed at the opening stage to engage in a civil disputation in which each party takes turns putting forth a reasonably short argument to support its view, and then allows the other party to criticize that view or to put forward arguments for the opposed view. In short, we assume that the discussion meets the requirements of a persuasion dialogue. We also assume that there are some other people present at the dinner, including the host, and that this group of people will be taken to represent the audience. Perhaps we can even assume that the host acts as a kind of moderator who assures that the parties take turns making contributions to the discussion, and each of them sticks to arguments that are relevant to the dispute they agreed to participate in at the opening stage. We also assume that, at some point, the host will close the discussion when he feels that it has gone on far enough to be interesting without becoming boring. In other respects, however, we will assume that the discussion is not regulated more strictly in the manner of a forensic dispute or a legal trial. For convenience, each of the moves is numbered in the dialogue on tipping below.

1. Helen: I'm against tipping because tipping contributes to low self-esteem and self-esteem is good.
2. Bob: But if someone is doing a better job, they should be paid better. And anyway, what evidence do you have that tipping contributes to low self-esteem?
3. Helen: Tipping makes the server feel undignified.
4. Bob: But tipping can be a good source of income and earning a good income makes a person feel secure and dignified. Tipping shows that in the customer's opinion the employee has given good service, and being perceived to have given good service rewards self-esteem. You admitted yourself that self-esteem is good.
5. Helen: Over the long run this practice of continual tipping leads to a lowering of self-esteem because continual tipping leaves a lingering feeling of being socially inferior. Because self-esteem is good, loss of self-esteem is bad. Tipping causes employees to lose self-esteem because tipping makes employees feel socially inferior, and feeling socially inferior leads to loss of self-esteem.
6. Bob: Tipping helps a business remain profitable that might be under financial stress. Also, it helps employees who may need the money to support their families.
7. Helen: Well, this may be true, but the problem is that this getting away without paying proper wages and so forth is very bad. All working men and women need to have proper wages and benefits.
8. Bob: Employers can still pay proper wages and benefits even though their employees are getting tips.
9. Helen: Employers won't do these things unless they have to. If they attempt to do these things they will make less income and it will put them under financial stress to survive in a competitive business.
10. Bob: Basically tipping is good, however, because it rewards excellence of service.
11. Helen: I don't think so, because a tip becomes an expected practice, whether the service is good or not.
12. Bob: If someone expects a tip for poor service, the client should correct this expectation. Customers should use their consumer skills.
13. Helen: One time a waiter intentionally spilled soup on my husband's new suit. Many misunderstandings of this kind happen

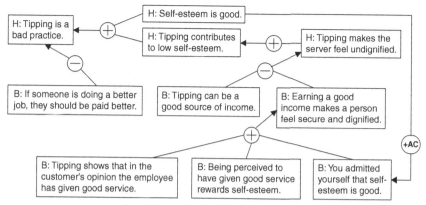

FIGURE 8.2. Moves 1–4 in the Dialogue on Tipping

in connection with tipping, and they can lead to all kinds of serious problems.

14. Bob: A lot of people would not get jobs if it weren't for tipping. Getting jobs is an important benefit during this time of high unemployment.

Instead of visualizing the whole sequence of fourteen moves as one very large argument diagram, we make the argumentation simpler for the reader to follow by breaking it down into four parts. The first part, shown in Figure 8.2, represents the first four moves in the dialogue on tipping. Helen's ultimate conclusion to be proved, the proposition that tipping is a bad practice, is shown at the top left of Figure 8.2. To the right of it, Helen's first argument is shown. It is a pro argument with two premises. One is that self-esteem is good and the other is that tipping contributes to low self-esteem. These premises are taken together as a linked pro argument supporting Helen's ultimate conclusion.

Next, Bob's first move is presented when he says that if someone is doing a better job, they should be paid better. This move is represented as a con argument against Helen's ultimate proposition to be proved. In other words, it is against Helen's thesis that tipping is a bad practice, and therefore we can infer that it could also be seen as a pro argument supporting Bob's thesis that tipping is a good practice.

In general, it doesn't matter which thesis, Bob's or Helen's, we represent as the ultimate thesis to be proved. We just have to adjust the pro and contra values accordingly. For example Bob's argument that if someone is doing a better job they should be paid better can be represented as in Figure 8.2 as a con argument against Helen's ultimate claim, or it could equally well be represented as a pro argument supporting Bob's ultimate claim that tipping is a good practice. We have chosen to represent the ultimate conclusion in the argumentation in the dialogue on tipping as Helen's claim that tipping is a bad practice. But we could have equally well done everything the other way around by representing the ultimate conclusion as Bob's claim and then simply change every pro to contra and vice versa.

Next we represent Helen's argument at move 3 when she says that tipping makes the server feel undignified. We take this as a pro argument for her previous claim that tipping contributes to low self-esteem. Hence, this latter proposition plays the role of a conclusion in one pro argument and the role of a premise in the next pro argument, making up a chain of arguments. Finally, we represent Bob's counterargument at move 4 against Helen's claim in her move 3 that tipping makes a server feel undignified. Bob's con argument has two premises, one is that tipping can be a good source of income, while the other is that earning a good income makes a person feel secure and dignified. This latter premise is then supported by Bob's pro argument shown at the bottom of Figure 8.2 with its three premises. Note that his third premise (his statement to Helen that she admitted herself that self-esteem is good), is shown as supported by Helen's previous statement (shown at the top of the diagram) asserting that self-esteem is good. This structure illustrates the feature of Carneades that a premise used in one argument can be reused in another one. The argument is shown on Figure 8.2 as fitting the scheme for argument from commitment, which basically says that if an arguer has gone on record as stating a proposition then she is committed to that proposition.

Next we turn to Helen's counterarguments at move 5. At this move Helen presents two arguments that are contra Bob's premise in his previous argument that being perceived to have given good service rewards self-esteem. In Figure 8.3, Bob's premise is shown at the left of the diagram, and the notation is included below it indicating that you have to track back to Figure 8.2 to find out where this premise occurs in the previous sequence of argumentation. Helen's first counterargument has two premises, shown at the top and middle of Figure 8.3. One of these premises is that loss of self-esteem is bad. Helen's argument at move 5 is that because self-esteem is good, loss of self-esteem is bad. To connect this to the previous line of argumentation, once again we have to track back to Figure 8.2 as indicated to the right of the premise that loss of self-esteem is bad in Figure 8.3. The reader might recall, that as shown in Figure 8.2, Helen had made the claim that self-esteem is good.

Helen's second counterargument at move 5, shown at the bottom of Figure 8.3, is her claim that tipping causes employees to lose self-esteem, backed up by the two premises shown to the right of the text box expressing this assertion. To sum up, what Helen has done at move 5 is to attack one of the premises in Bob's previous argument.

The next step is to analyze the sequence of argumentation going from moves 6 to 9. What is of interest here is that instead of attacking Helen's

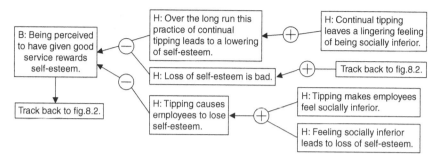

FIGURE 8.3. Helen's Arguments at Move 5

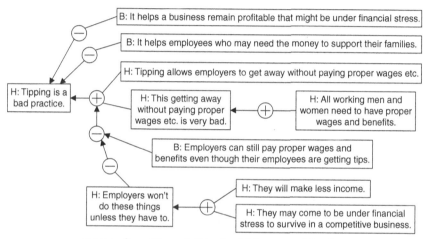

FIGURE 8.4. Moves 6–9 in the Dialogue on Tipping

previous argument by bringing forward con arguments against it, Bob starts a new line of argument. He argues that tipping helps a business remain profitable that might be under financial stress, and along with this claim he adds another claim stating that tipping helps employees who may need the money to support their families. The problem is to determine what the conclusion of this argument is. It appears that the conclusion of this argument is Bob's ultimate claim to be proved in the dialogue as a whole, namely the proposition that tipping is a good practice. So because we are taking Helen's thesis, the proposition that tipping is a bad practice, as the root of the tree, we will interpret Bob's new argument as being directed against Helen's ultimate claim. So if we look at Figure 8.4 at the top, we see that these two propositions put forward by Bob are represented as contra arguments against Helen's thesis that tipping is a bad practice.

At move 7, Helen concedes Bob's argument by saying "Well, this may be true," and then she goes on to make a statement about what she thinks the problem is. We take it that what she is doing here is presenting a new line of argument to support her conclusion that tipping is a bad practice. First she says that this getting away without paying proper wages is very bad, and then she backs it up with the second statement saying that all men and women need to have proper wages and benefits. This argument can be taken to be put forward in support of Helen's ultimate claim that tipping is a bad practice, and so that is the way it is represented in Figure 8.4.

Next, at move 8 Bob reacts to Helen's last argument made at move 7. He says that employers can still pay proper wages and benefits even though their employees are getting tips. We interpret this move as an undercutter attacking Helen's previous argument. Then at move 9, Helen brings forward a new argument directed against Bob's previous argument. In her new argument, Helen states that employers won't do these things unless they have to, and she backs up this claim with two additional premises. These three statements, as shown in Figure 8.4, make up a counterargument that attacks Bob's previous counterargument against Helen's previous argument. So what we have here is a counterargument to a counterargument to a previous pro argument. This nested sequence represents a chain of arguments in which each one attacks the next one, characteristic of abstract argumentation frameworks of the kind used by Prakken and Sartor to model burden of proof.

Finally, we only need to examine the structure of the argumentation in the remaining moves, moves 10 to 14. Once again, instead of attacking Helen's previous argument in the dialogue, Bob chooses to revert back to offering evidence to support his ultimate claim that tipping is good. This is represented in Figure 8.5, at the top of the figure, by Bob's contra argument against the ultimate conclusion shown at the left that tipping is a bad practice. Bob's line of argument is shown in Figure 8.5 along the top of the argument diagram where he makes the claim that tipping rewards excellence of service. Helen now attacks Bob's argument with her counterargument claiming that it becomes an expected practice whether the service is good or not. Bob then, in turn, attacks Helen's counterargument with another counterargument stating that if someone expects a tip for poor service, the client should correct this expectation, and also stating that customers should use their consumer skills. These two premises seem to go together, and so we have interpreted them as acting as two premises in the linked counterargument against Helen's previous claim.

At move 13, Helen introduces a new argument by citing an incident in which a waiter intentionally spilled soup on her husband's new suit. She claims that many misunderstandings of this kind happen, and that they lead to serious problems. This argument can be represented as an instance of argument from negative consequences to the effect that tipping is a bad practice because it has such negative consequences. Bob then attacks Helen's argument from negative consequences using an argument from

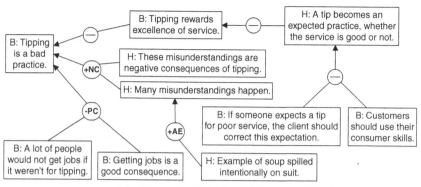

FIGURE 8.5. Moves 10–14 in the Dialogue on Tipping

positive consequences as a con argument against her previous argument. In other words, Bob's argument is being used as an undercutter to attack Helen's previous argument. NC represents argument from negative consequences, PC represents argument from positive consequences and AE represents argument from example.

Now we can connect all the arguments displayed in Figures 8.2 through 8.5 together as indicated, and the result is a graphical representation in the Carneades Argumentation System of the whole chain of argumentation woven throughout the dialogue on tipping.

5. Burdens of Proof in the Dialogue on Tipping

We are assuming that in the dialogue on tipping there is no problem about determining the burden of persuasion. According to the description of the dialogue given in Section 4, it is clear to all the participants in the dinner party that Helen is against tipping and Bob is for it. Moreover the host, who has agreed to act as moderator for the discussion between Helen and Bob on tipping, has made it clear to all the parties that Helen's ultimate conclusion to be proved is the proposition that tipping is a bad practice and Bob's ultimate conclusion to be proved is the proposition that tipping is not a bad practice. Thus, Helen's burden of persuasion is to prove her ultimate proposition and Bob's burden of persuasion is to prove his ultimate proposition. The problem is to determine whether the notions of evidential burden and tactical burden can be applied to the argumentation in the dialogue on tipping.

This kind of case is different from a case of legal argumentation in a trial setting because in a trial the trier (judge or jury) is taken to be the audience who determines the outcome of the argumentation in the case. Moreover, in a legal setting there are specific rules about what kinds of evidence are admissible, and about which arguments are judged to be irrelevant and excluded from the trier's consideration by the judge. Another factor is that in law, the evidential burden, also called the risk of failure to bring forward evidence, is decided by the judge. In law this burden is a special procedural measure that enables the trial to come to an outcome under conditions where resources (time and money) are limited. If one side has presented strong enough evidence to meet its present burden of persuasion, and the other side has not tried to undermine this evidence or even question it, the judge can simply declare the trial over, on the basis that the second side has failed to produce the evidence needed to support its side of the case, and that therefore the trial has reached the closing stage.

It is uncertain whether this burden of production of evidence is applicable in everyday conversational exchanges or in debates. What is important is that there is some second kind of burden, different from the burden of persuasion, that shifts back and forth as argumentative moves are made over the sequence of a dialogue even though the burden of persuasion has been set at the opening stage. It doesn't matter very much for our purposes here whether we think of the second burden as an evidential burden or a tactical burden. It functions in both ways.

The main problem with evaluating evidential and tactical burdens of proof in the argumentation in the dialogue on tipping depends on who the audience is taken to be, and which propositions can be taken to be accepted by the audience. To get some idea of how to go about making such determinations as applied to a particular example, let's go back to considering the first few moves by each side in the dialogue on tipping. Helen's first argument was based on her two premises that self-esteem is good and that tipping contributes to low self-esteem. The second premise is supported by an additional argument with the single premise that tipping makes the server feel undignified. These two premises, along with the additional argument to support one of them, are used in her argument to support her ultimate conclusion that tipping is a bad practice. How could we expect the audience to react to Helen's argument? Let's say they react as shown in Figure 8.6. AC represents argument from commitment.

Let's say that the audience accepts Helen's statement that self-esteem is good. There are grounds for doubting this claim. For example, some might feel that too much self-esteem is a bad thing, and might therefore have doubts about the claim that self-esteem is unconditionally good. But generally speaking, there is widespread acceptance of the view that self-esteem is a good thing. Let's say then that the audience accepts this proposition.

Next, let's look at Helen's statement that tipping contributes to low self-esteem. This seems to be a kind of empirical claim one might find in the social sciences. It seems reasonable enough, but an audience might demand some evidential support for it before they would accept it. To provide this support Helen has offered the premise that tipping makes the server feel

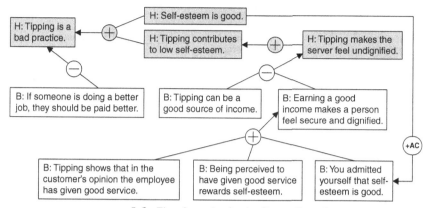

FIGURE 8.6. First Steps in the Audience Reaction

undignified. Once again, this seems to be a kind of empirical statement about behavior, but let's say the audience is willing to accept this proposition without any further proof, at least for the time being. At the stage represented by the argument diagram shown in Figure 8.6 then, without considering any further evidence or argumentation, Helen has fulfilled her burden of persuasion.

Next, we have to take into account the two con arguments presented by Bob. We need to look to make some estimate of whether the premises in these arguments would be acceptable to the audience or not. Bob's one argument is based on the premise that if someone is doing a better job they should be paid better. A lot of people would accept this proposition, but it is politically sensitive, and we can reasonably assume that a lot of people might be uncomfortable with it as a general principle, and would be unwilling to accept it. So if we look at the next step in the argumentation represented in Figure 8.7, this premise is represented as stated but not accepted.

Bob's second argument against Helen's statement that tipping makes the server feel undignified seems to be stronger. The three premises shown in the text boxes along the bottom have been shown in darkened text boxes, indicating that the audience at the dinner party might be expected to find them acceptable. As noted earlier, Bob even used argument from commitment to support the premise that Helen admitted that self-esteem is good by tracking it back to her earlier statement that tipping is good, shown at the right of Figure 8.7.

Given that all three premises of Bob's argument can be taken to be accepted by the audience, we can show that the conclusion of this argument, the statement earning a good income makes a person feel secure and dignified, would also be accepted by the audience. This statement in general seems quite reasonable and would likely be broadly accepted in the kind of audience one might expect, so it doesn't require a very high

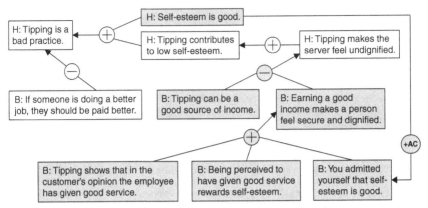

FIGURE 8.7. Next Steps in the Audience Reaction

standard of proof in order to present a successful proof for it to make this audience accept it, if they didn't accept it beforehand. So let's say, on balance, that the audience accepts this proposition, based on the argument given to support it. The other premise and Bob's next argument, the statement that tipping can be a good source of income, is an empirical proposition, one that few people would doubt as an item of common knowledge.

Now having tracked back the argument this far, as shown in Figure 8.7, we see that Bob's counterargument against Helen's claim that tipping makes the server feel undignified is fairly strong. Also, Helen's claim that tipping makes the server feel undignified is not backed up by any evidence that Helen has offered. So although it was acceptable before it was challenged by Bob, Bob's counterargument is strong enough to cast it into doubt. Therefore, we now represent Helen's statement that tipping makes the server undignified as stated but not accepted. It follows from this that Helen's statement that tipping contributes to low self-esteem, which was based only on the evidence of her statement that tipping makes the server feel undignified, is now also marked as stated but not accepted. The upshot of this whole sequence of argumentation is that Helen's ultimate conclusion, her statement that tipping is a bad practice, is no longer accepted. In other words, the evidential burden of proof, and along with it the tactical burden of proof, has shifted from Bob to Helen. At this point, if Helen fails to produce any new arguments to support her ultimate claim that tipping is a bad practice, or if she fails to defeat any of Bob's arguments that were just put forward, Bob wins the disputation.

However, this tracking of the shifting of the burden of proof has only gone through moves 1 to 4 in the dialogue on tipping. The evidential burden of proof shifts against Bob's side as Helen introduces her two counterarguments in move 5 because Helen backs up each of these arguments with supporting premises that would be fairly plausible to the audience unless they had specific reasons to doubt them. To get an idea of how the evidential and tactical burdens shift beyond that point, we have to track through the sequence of argumentation in all of moves 6–14 of the dialogue. To start this procedure we have to look back to Figure 8.4.

The Carneades Argumentation System allows four possibilities: a proposition can be accepted, rejected or it can fall into the category of being stated but not accepted. Here however, for the sake of simplicity, we have only been allowing for two possibilities, acceptance and nonacceptance (in or out). Each proposition has the burden of proof depending on whether it fits into common knowledge, goes against common knowledge and generally how plausible it seems to be, meaning how willing or reluctant the audience would be to accept it, as far as we know, about what the audience would accept or reject. If we look through the propositions represented in the text boxes in Figure 8.4, they all seem fairly reasonable, except for the ultimate conclusion that is the topic of the debate. If we look at the top two arguments in Figure 8.4 put forward by Bob, they seem like they might be acceptable to the audience, even though they could be questionable. And below these two arguments we have a pro argument put forward by Helen with two premises, and one premise is supported by another claim stating that all working men and women need to have proper wages and benefits. This claim is questionable, but it seems likely that the audience might accept it. So far then, it seems we have a tie. We have two contra arguments put forward by Bob and one pro argument put forward by Helen, and one of the premises in Helen's argument is supported by another argument backing it up. But then at the bottom of Figure 8.4 we have two undercutters. The first one is put forward by Bob. Let's say that Bob's argument undercuts Helen's argument just above it about paying proper wages and benefits. Now let's look to the bottom of Figure 8.4 where Helen undercuts by making the claim that employers won't do these things unless they have to. She supports this claim with an argument having two fairly plausible premises that the audience might be inclined to accept. So at first there was a tie between the arguments of Helen and Bob, and he undercut her argument, but then she, in turn, undercut his argument. This outcome leaves Bob victorious, as he still has the two arguments at the top and neither is undercut by any argument of Helen. Of course she could still attack these two arguments when it is her turn, but she does not.

At this point, Carneades is not as decisive as it could be ideally because we are not sure that the audience really accepts all the statements that have been put forward by both sides. No doubt some of the people in the audience accept some of the statements in question or reject others, while others in the audience have different opinions. So, as far as we are told in the dialogue on tipping, there is very little specific information on what the particular audience in the example accepts or does not accept. The best we have been able to do is to make some fairly loose assumptions about what they might be inclined to accept or not. Still, this does not mean that the model is deficient, or is not helping us, because in many instances of everyday conversational argumentation, although we can definitely say with respect to some propositions that the audience being addressed would accept them or not, in many other instances there are lots of propositions that we simply don't know about in this respect. We can always ask the audience, or find some other way of trying to determine acceptance or rejection, but we will have to expect that in many instances this is not possible or practical.

At any rate, it now looks like Bob could be ahead in the disputation, given the sequence of undercutters in Figure 8.4 at the bottom, so let's go ahead to moves 10 to 14 in the dialogue as represented in Figure 8.5. Here Bob puts forward the con argument that tipping rewards excellence of service, as shown at the top of Figure 8.5. We could assume that the statement that tipping rewards excellence of service would be acceptable to the audience, but Helen counterattacks with a contra argument by claiming that it becomes an expected practice whether the service is good or not. But then Bob attacks this premise with a counterargument based on two premises: (1) the premise that if someone expects a tip for poor service, the client should correct this expectation, and (2) the premise stating that customers should use their consumer skills. If we assume that the audience accepts these two premises then Bob has defeated Helen's counterargument and so his argument based on the premise of excellence of service is reinstated. This evidential situation shifts the burden of proof to Helen's side. Can Helen meet this evidential burden with further arguments?

Helen has a graphically powerful argument from negative consequences ultimately based on her example of soup having been intentionally spilled on her husband's suit. She uses this argument to support her ultimate conclusion that tipping is a bad practice. So far then, we have a pro argument and a contra argument both leading to Helen's ultimate thesis that tipping is a bad practice. But then we have to take into account Bob's contra argument against Helen's ultimate conclusion that tipping is a bad practice. This argument from positive consequences is shown at the bottom left of Figure 8.5. If the audience accepts both premises of this argument, and they both seem reasonably plausible even though they are not supported by any further evidence, then this argument counterbalances Helen's pro argument from negative consequences. Thus, it would seem that at this point the argument is a stalemate. First Bob put forward a contra argument that was attacked by Helen's counterargument that was then attacked by Bob's contra argument. Then each party has put forward an argument from consequences, and these two arguments can be seen as cancelling each other out. The upshot is that Helen has not met her evidential burden with her pro argument from negative consequences, assuming that it is matched by Bob's con argument from positive consequences. So Bob is still slightly ahead, unless Helen can counter Bob's argument represented at the top of Figure 8.5 stating that tipping rewards excellence of service. This might not be too hard for Helen to do. For example, she could attack Bob's argument that customers should use their consumer skills when someone expects a tip for poor service. She could argue, for example, that this is not effectively possible in many instances, and that it could lead to disputes that make people uncomfortable, or even lose their self-esteem.

Having tracked over the whole sequence of argumentation running through moves 10 to 14 in the dialogue, what our analysis suggests is that the evidential, and the tactical burden of proof as well, have shifted back and forth numerous times as new arguments are put forward by one side or the other at succeeding moves in the dialogue. Our tracking system for determining the precise locations of these shifts of the burden of proof is imperfect because our judgments of which propositions the audience accepts or rejects in this particular example are not exactly known, and we can only make reasonable assumptions in some instances about how the participants and any observers should make such determinations. But even so, the method has enabled us to determine, based on the information that we do have, where the evidential and tactical burdens of proof shift from one side to the other. The more knowledge we have about what a given audience accepts or not, the more precise determination we can make about where these burden of proof shifts should occur. Understandably, we will not know as much about the audience as we do in a court of law, where the trier has to make a decision based both on the law and the facts of the case, and many of the facts of the case are known by the evidence introduced in the trial. We might also know more in a carefully regulated forensic dispute where some factual information is agreed on by both parties at the opening stage of the dialogue.

6. Burdens of Proof in a Forensic Debate

First of all, we start out with a type of dialogue. It could be any type of dialogue, but let's say it is a persuasion dialogue. Let's say there are three parties, the proponent, the respondent and an audience. The burden of proof is set at the opening stage. There can be various ways of doing this. One type of situation is that only the proponent has a burden of proof. In other words, the proponent wins the dialogue if she produces a chain of argumentation strong enough to prove her thesis initially set to be proved at the opening stage. The task of the respondent is merely to cast enough doubt on the proponent's attempts to prove her thesis so that she ultimately fails to prove it. What constitutes a successful proof depends on the standard of proof set for the proponent to achieve using her argumentation. The standard can be higher or lower. Another possible situation is that the proponent is designated the task at the opening stage of proving her thesis T, while the respondent is designated the task of proving his thesis ~T. These situations represent two types of opposition. The second kind of situation represents a dispute type of dialogue, where both parties have a thesis to prove, and the thesis of the one is the opposite of the thesis of the other. The first type of situation represents a dissent type of dialogue, where only the one side has the burden of proof and the task of the other side is the negative one of attacking the arguments of the first side to show weaknesses in them. The role of the second side is that of a critical questioner.

There can be various ways of setting the burden of proof at the opening stage, depending on the standard of proof set for each side. For example, in one very common kind of case, the preponderance of evidence standard may be set for each side. Once the standard is set, it means that at the closing stage of the dialogue, if one side produces a chain of argumentation that supports its thesis even very slightly more strongly than the support given by the argumentation of the other side for its thesis, then the first side wins. In other words, one side, in order to beat the other, has to only produce an argument that is slightly stronger than that produced by the other. This way of setting the burden of proof at the opening stage rests on a balance of probabilities. Whichever side produces a chain of argumentation that supports its conclusion with a probability value of greater than .5 is the one that is judged to be successful in the dialogue.

There can be many other ways of setting the burden of proof at the opening stage. Another way that might be appropriate for a dialogue that represents a debate between two arguers to try to persuade an audience is to set a burden of persuasion ruling that whichever side has a more persuasive impact on the audience wins the dialogue. This way of setting burden of proof requires that the audience already has an assigned degree of acceptance for the thesis of each side at the opening stage. Let's call this the initial plausibility value of each thesis. As the dialogue proceeds, each side tries to increase the plausibility value of its own thesis in the eyes of the audience. In Figure 8.8, $T(P)$ is the proponent's thesis and $T(R)$ is the respondent's thesis. We assign plausibility values u, x, y, z,... to each thesis at each stage of the dialogue assuming that the values for x and y are both greater than zero at the opening stage, and that these values can increase or decrease as the argumentation proceeds. As shown in the left column in Figure 8.8, the initial value of the proponent's thesis (at the opening stage) is designated by u, and the final value of it (at the closing stage) is designated by x. As shown in the right column, the initial value of the respondent's thesis (at the opening stage) is designated by y, and the final value of it (at the closing stage) is designated by z. The arrows represent the sequences of moves that take place during the argumentation stage of the dialogue. The burden of persuasion for each participant can be expressed as follows.

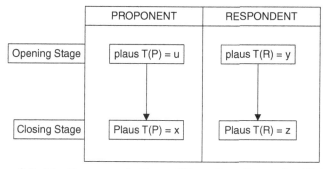

	PROPONENT	RESPONDENT
Opening Stage	plaus T(P) = u	plaus T(R) = y
Closing Stage	Plaus T(P) = x	Plaus T(R) = z

FIGURE 8.8. The Structure of a Type of Three-Party Persuasion Dialogue

Proponent: $(x - u) > (z - y)$
Respondent: $(z - y) > (x - u)$

If at the closing stage we have $(x - u) = (z - y)$, the outcome of the dialogue is a tie. Neither party has succeeded in fulfilling its burden of persuasion.

The structure of this debate type of dialogue is that of a dispute. If the proponent's burden of persuasion is met, the respondent's burden of persuasion is not, and vice versa.

To take an example, let's say that at the beginning of the dialogue the audience finds the thesis of the proponent highly plausible. Let's say that the audience would assign it a very high value of .9. However, the audience does not find the thesis of the respondent very plausible at all. Let's say that the audience would assign it a very low plausibility value of .2. On this basis, we can judge the persuasive success of each party by the degree to which they increase the plausibility value of their thesis by the argumentation that has been put forward in the argumentation stage. Let's say that the probability value of the proponent's thesis only went up to .92, at the closing stage, whereas the probability value of the respondent's thesis went up to .5. In this case, the respondent wins because the audience found his argument to increase the plausibility value of his thesis by .3, whereas they only found the proponent's argumentation to increase the plausibility of her thesis by a value of .02. Thus, even though the audience found the thesis of the proponent much more plausible at the end of the dialogue, it is the increase of the plausibility value of his thesis that occurred over the course of the dialogue, compared to that of the opposing side, that determines which side was more successful.

So far we have only considered what happens at the opening and closing stage where the burden of persuasion is brought into play to determine which side wins. However, we can also look at how the plausibility values of the audience are affected during the argumentation stage. As the dialogue proceeds into the argumentation stage, the audience evaluates each argument put forward at each move, and perhaps also at subsequent moves where that argument is questioned or criticized. This is where burden of production comes in. Let's say that at some particular move in the dialogue the proponent puts forward a very strong argument that increases the plausibility of her thesis to such a degree that if the argumentation were to be terminated at that point, she would clearly win the dialogue. The proponent has therefore moved forward with producing evidence, putting the respondent on the defensive. It means that if the audience were to make up its mind at that particular point in the dialogue, they would clearly accept the proponent's thesis. But if the dialogue is not at the closing stage yet, this status could still change if the respondent successfully attacks her argument, or brings forward other arguments to support his thesis.

It is this kind of situation that is often associated with the shift in the burden of proof. What it means is that there has been a shift in initiative. Perhaps the argumentation was equally balanced before, or the audience was even more persuaded by the respondent's previous arguments. However, now the proponent has come forward with this strong argument, one which the audience accepts as highly plausible, the proponent has seized the initiative. If he fails to reply in a strong enough way by producing more arguments in the time that is left, she will prevail. Before, the initiative may have been balanced, or it may have been on the respondent's side, but it has now shifted to the proponent's side. If the respondent doesn't do anything further at this point, he loses the dialogue. This species of burden of proof is called the burden of production, or the burden of producing evidence. Sometimes it is also called the burden of going forward with evidence. All these terms are appropriate. They are also useful, because they enable us to distinguish between burden of production and burden of persuasion, thus helping to disambiguate the terminology commonly used to indicate or describe burden of proof.

While the burden of production is judged by the audience, the proponent and the respondent often need to make an assessment themselves at various moves of the dialogue about how successful their argumentation has been so far and what they need to do in the future part of the dialogue to win the outcome at the closing stage. This notion is the tactical burden of proof. Let's go back to the same example just earlier where the proponent has put forward a strong argument that the respondent needs to counter.

This case is theoretically interesting, for a number of reasons, because it represents the midpoint between everyday conversational argumentation and legal argumentation in a trial, where the judge applies legal rules and facts, and acts as a third-party audience. In everyday conversational argumentation, in many of the most common instances at any rate, there are only two parties to the discussion, and these two parties have generally not made specific agreements at the opening stage on what they will accept as evidence, and that sort of thing. The example on tipping is the case in point. It is also theoretically interesting because it depends on the assumption that how strongly these statements are accepted or not accepted by the audience can be represented numerically as a fraction between zero and one. This scale of measurement is characteristic of the probability calculus. However, it is extremely controversial whether numbers of this kind can be attached to the propositions making up the premises and conclusions in a chain of argumentation in a legal trial setting. Both the analyses of legal burden of proof by Prakken and Sartor and in the Carneades Argumentation System operate on the assumption that it is not realistically possible to make such numerical assignments in the normal run of legal cases. Of course, the same kind of objection may be very likely to be put forward by those in the field of speech communication who study argumentation in forensic debates. If so, the kind of structure represented in Figure 8.8 would not be applicable to modeling burden of proof in that setting. And if this is the case, they might find the approach taken by Prakken and Sartor and by the Carneades Argumentation System more useful.

7. The Connection between Burden of Proof and Presumption

How is presumption different from ordinary defeasible inference? As shown in Chapter 3, a legal presumption looks to have the same structure as an ordinary defeasible inference. In the most typical cases, as we saw in Chapter 3, a presumption is based on two premises, called the fact and the rule. The fact can be described as an atomic proposition in logic, a simple statement that is not conditional (disjunctive, conjunctive) in form. It is called a "presumption-raising fact" in law, and that terminology can be retained here. In law, the facts of a case consist of the evidence judged to be admissible at the opening stage of a trial. A fact is a judicially admitted proposition. The rule is often described as a generalization. Rules can be defined by the seven characteristics from (Gordon, 2008, 4) cited in Chapter 3, Section 3, indicating that rules are subject to exception, can conflict and can become valid or invalid. This notion of a rule cannot be modeled adequately by material implication of the kind used in deductive logic. Instead, rules need to be modeled by identifying the parts of the rule – antecedent, consequent, exceptions, assumptions and type, as done in the Carneades Argumentation System.

There are different theories about which part of a defeasible inference is to be identified as the presumption. On one theory, the presumption is to be identified with the defeasible rule (Prakken and Sartor, 2006). Often the presumption is identified with the conclusion. It is said that the fact and the rule together "give rise to" the presumption stated in the conclusion. Still, other writers talk about presumptive reasoning by equating the presumption with the whole inferential process leading from the fact and rule to the conclusion drawn from it. However, although presumption may be correctly identified as a defeasible inference of this kind with two premises and a conclusion, there is another question to be raised. What is the difference between an inference and a presumption?

What makes presumption different from other kinds of inferences is that it is put forward in a special way in a context of dialogue where two parties are reasoning together, and a global burden of proof has been set at the opening stage. When one party puts forward an assertion or argument to the other party in such a context, the assertion or argument is typically put forward in such a way that the other party is meant to either accept the assertion or argument or challenge it in some way. It can be challenged by raising doubts about it by asking critical questions, or by demanding some proof of what has been asserted. The respondent to the assertion or argument presented normally has such a right of challenge. The proponent's responsibility to provide such a proof (to the required standard of proof) is called the burden of proof. What makes presumption different as a way of putting forward a proposition for acceptance in a dialogue is that this right of the challenge is temporarily removed. When a presumption is put forward there is a shift. Instead of there being a burden of proof on the side of the proponent, the burden may shift to the other side to disprove the proposition in question.

Another distinguishing factor that makes a presumption different from an inference that is not presumptive in nature is the probative weight of the premise stating the rule. Normally when an inference is put forward in the form of an argument, the proponent of the argument has to support the premises if any of them are challenged by the respondent. A premise supported by evidence is said to have probative weight. It is this probative weight that moves the argument forward as a device that forces the respondent to accept the conclusion given that the argument has a valid logical form. However, in the case of a presumptive inference, a problem is that there is insufficient evidence to prove the premises and give them enough probative weight to move the argument forward toward acceptance. What fills the gap in the case of presumptive inference is that one premise is a rule that is accepted by procedural reasons even though it lacks the probative weight that would be bestowed upon it by sufficient evidence. In law, the distinction is drawn as follows: "[An] inference arises only from the *probative force of the evidence*, while the "presumption" arises from the rule of law" (Whinery, 2001, 554). More generally, a presumption arises from a rule that is established for procedural and/or practical purposes in a type of rule-governed dialogue (like a trial).

Global burden of proof in a dialogue is defined as a set $<P, T, S>$ where P is a set of participants, T is an ultimate *probandum*, a proposition to be proved or cast into doubt by a designated participant and S is the standard of proof required to make a proof successful. If there is no thesis to be proved or cast into doubt in a dialogue, there is no burden of proof in that dialogue, except where it may enter by a dialectical shift. The local burden of proof defines what requirement of proof has to be fulfilled for a speech act, or move like making a claim, during the argumentation stage. The global burden of proof is set at the opening stage, but during the argumentation stage, as particular arguments are put forward and replied to, there is a local burden of proof for each argument that can change. This local burden of proof can shift from one side to the other during the argumentation stage as arguments are put forward and critically questioned. Once the argumentation has reached the closing stage, the outcome is determined by judging whether one side or the other has met its global burden of proof, according to the requirements set at the opening stage.

Let's take the example of an inquiry dialogue as the best starting point for illustrating the connection between presumption and burden of proof. The global burden of proof for an inquiry is set at the opening stage. For the proposition that is unsettled to be proved so that the dialogue can be closed, this proposition needs to be proved to such a high standard that there will be no need to retract it in the future. Although such a high standard is the one to be aimed for, normatively speaking, in practical terms retraction is a reality that in many cases cannot be ruled out. For example, in a scientific inquiry, let's say in theoretical physics, some theory may be proved to very high scientific standard of proof, but that is no guarantee that a new theory will not come along in the future that shows that the previous theory needs to be retracted, either generally or as applied to specific phenomena. There may just not be enough evidence available at this point to prove the theory to such a high standard of proof that it can be guaranteed that it will never be retracted. After all, scientific argumentation is by its nature defeasible. How can a scientific investigation move forward in light of this problem? The answer is that it will have to rely on presumptions.

In a scientific investigation, presumptions take the form of statements that have some evidence supporting them, that generally seem reasonable in light of accepted scientific findings and that have no known scientific evidence that goes against them. They can be provisionally accepted as hypotheses, subject to later retraction if fault is found with them, as premises needed to drive the line of argumentation forward that aims at the ultimate proposition to be proved or disproved. Thus, a proposition may be taken as a presumption or acceptable hypothesis in an inquiry if a good deal of evidence supports it, and no known evidence goes against it, even though that weight of evidence is not strong enough to justify it as part of the chain of argumentation culminating in a proof that meets the standard of proof appropriate for the inquiry. Such a hypothesis may be a weak link in the chain of argumentation, but further evidence may be found to support it in the future and then its evidential weight may be boosted up enough to meet the required standard of proof. But support for the hypothesis may be too weak presently to meet that standard, even though its provisional acceptance is a required step in the sequence of argumentation leading to the ultimate conclusion to be proved. In such a case, its acceptance needs to be justified on the basis that the hypothesis is a presumption.

When talking about presumptions, both in law and everyday conversational argumentation, the aspect typically called the shifting of the burden of proof is described as follows. When an assertion in an argument is put forward, a proponent has the burden of proof to support it with evidence if it is challenged by the respondent. When a presumption is put forward, however, this burden of proof on the respondent may be removed. The presumption is put forward as a proposition that the respondent has to accept. He can't demand proof of a kind that would normally be required to back it up. It is as if the presumption has to be accepted as a stipulation. Reasons can be given to back up acceptance of the presumption, but they are typically practical reasons relating to the continuation of the dialogue that is underway, as opposed to evidential reasons of the kind one would normally use to back up or prove a claim made.

So here we see the relationship between global burden of proof and presumption. The global burden of proof may be set to too high a standard to be reasonably met by the given evidence. In such a case the inquiry will fail, and the conclusion must be drawn that we can't prove the ultimate proposition to be proved, and therefore we must conclude that it can't be proved. But such a conclusion is based on epistemic closure, that is, on the assumption that all the evidence has been collected so that the inquiry can be closed. However, this way of proceeding represents a normative ideal that may be in practice impossible to achieve because of the fluidity of the evidence. It may be that the proof on the given evidence is fairly conclusive, but that a few of the links in the chain of argumentation are weaker than they should be. Still, the evidence on the whole may be strong enough to justify drawing the conclusion, so that subject to certain specific reservations, the unsettled proposition that is the aim of the inquiry may be said to be proved. A simpler way to accomplish the same end would be to simply lower the standard of proof. But for various reasons, that may not be something that is desirable. An alternative is to accept the proposition as a premise that does not meet the standard of proof required to properly accept it, but that has enough evidence supporting it so that it can be accepted as a presumption.

In short, the notion of presumption needs to be defined in terms of the notion of standard of proof, the notion that is built into the global burden of proof set for a dialogue. That is how the notions of presumption and burden of proof are connected. This is still a simplified explanation of the connections, however, because it only takes into account global burden of proof so far. A fuller explanation needs to take into account the connections between presumption and two other most central types of burden of proof, the burden of going forward with evidence and the tactical burden of proof. However, it is easy to see that the evidential burden, or burden of going forward with evidence, is closely connected to the notion of presumption. For it is by means of the application of the notion of presumption that it is possible to fulfill a burden of going forward with evidence in the sequence of argumentation where otherwise that would be impossible or impractical because of the lack of evidence.

The function of making a presumption is to enable a discussion or investigation to move forward without getting continually bogged down by having to prove a proposition needed as part of an argument required to help the investigation move forward. The problem may be that proving such a proposition may be too costly, or may even require stopping the ongoing discussion or investigation temporarily so that more evidence can be collected and examined. The problem is that a particular proposition may be necessary as a premise in a proponent's argument he has put forward, but the evidence that he has at present may be insufficient to prove it to the level required to make it acceptable to all parties. Hence, moving forward with the argumentation may be blocked while the opponent demands proof. The two parties may then become locked into an evidential burden of proof dispute where one says "you prove it" and the other says "you disprove it." This interlude may block the ongoing discussion. A way to solve the problem is for the proponent or a third party to say, "Let's let this proposition hold temporarily as a premise in the proponent's argument, so that we can say he has proved his contention well enough so that we can accept the conclusion of his argument tentatively as a basis for proceeding." If necessary, later on, the subdiscussion can be continued by bringing in more evidence for or against the proposition that served as the premise.

8. Dialectical Refinements of the Theory of Presumption

There is also a more subtle but no less important distinction to be drawn between a presumption and a putting forward of that presumption. The putting forward of a presumption can be seen as a kind of speech act in a dialogue, while the presumption itself can be identified, as indicated earlier, by the inference it is part of. The same ambiguity attaches to the concept of an argument, and is a common source of confusion. A distinction needs to be drawn between an argument, and the putting forward of an argument for acceptance in a dialogue. From one point of view, a traditional one in logic, an argument can be viewed simply as an inference from premises to a conclusion. Or, from another point of view, an argument can be seen as something that is put forward by one party for acceptance by another party. An argument, on this latter view, is something that is advanced or advocated by a claimant. It is something that has the function of backing up a claim by giving reasons to accept it.

Krabbe (2001) studied the problem of retraction in persuasion dialogue, and showed how the notion of a presumption is important for solving this problem (151–153). He offered an example of a dialogue (152) similar to the following one illustrating some conditions for retraction of a presumption. The dialogue illustrates a presumption in favor of a source of evidence that is generally accepted as trustworthy, like a weather forecast.

Wilma: The fine skating weather is holding.
Bruce: Why?
Wilma: The weather forecast says so.
Bruce: So what?
Wilma: You can usually trust the weather forecast. Why not in this case?

At his second move, Bruce refuses to accept Wilma's argument that the fine skating weather is holding because the weather forecast says so. When he says "So what?" he implies that he does not accept the weather forecast as a reliable source of evidence about the weather. But the problem is that he has given no reason why the weather forecast should not be accepted as a reliable source of evidence. Wilma replies at her last move by pointing out that the weather forecast is generally accepted as trustworthy. Here she is actually giving a reason to support acceptance of the inference that what the weather forecast said implies that the fine skating weather is holding. Whately (1846) would have analyzed this case by calling this acceptance a presumption in favor of authority. In more contemporary terms, we could say there is generally a presumption in favor of expert opinion.

As part of her last move, Wilma adds the remark, "Why not in this case?" at the end of her last move. This remark has the effect so often described as that of reversing the burden of proof. It is reminiscent of the recent literature on what should be the effect of asking a critical question in response to a defeasible argument like argument from expert opinion. In some instances, the asking of the critical question needs to be backed up by supporting evidence before the question defeats the original argument. Krabbe (151) makes this point by writing that after Wilma's last remark, it is up to Bruce to justify his challenging of the presumption that you can usually trust the weather forecast. Krabbe concludes, "Hence there has been a role reversal" for at that point in the dialogue, the burden of proof has fallen on Bruce, not Wilma (151). Krabbe uses this dialogue to make the point that even though presumptions may not be easy to retract, they are retractable, and need to be retracted under the right conditions in a dialogue structure that represents rational argumentation. Judging by this example it appears that Krabbe basically accepts the contentions of the Walton theory that one of the conditions under which a presumption needs to be retracted is that evidence is given against it, but that in a case like the example dialogue earlier where no such evidence has been given, the presumption stays in place.

However, it is evident from Krabbe's discussion that he sees the notion of a presumption in a different way from the way it is seen in the Walton theory. This difference is made evident in a remark in a footnote in Krabbe (2001, 158): "Walton stresses the way a presumption is introduced into the dialogue by a speech act of presumption. At present we are more interested in the way a presumption may be withdrawn from the dialogue." Following up this remark in a personal communication (e-mail of April 4, 2008), Krabbe wrote that Walton writes about presumption as a kind of speech act whereas Krabbe treats it as a kind of commitment. He added that what Walton calls a presumption he would call "proposing a presumption." It is difficult to compare the separate writings of Walton and Krabbe on presumption, even though both are based on a dialogue theory approach because there seems to be a basic terminological difference underlying the treatment of presumption in the two sets of writings. These matters take us back to the no commitment problem and the stability adjustment problem taken up in Chapter 5, Section 1. These problems are general ones for the dialogue model presented in Chapter 4.

These observations suggest the usefulness of drawing a distinction between two notions that are often confused: (a) the notion of presumption itself and (b) the speech act of putting forward a presumption for acceptance by another party in a dialogue. This distinction is fundamental and highly important, despite the fact that it has not been clearly recognized in the past and is often overlooked. Interestingly, the same kind of fundamental ambiguity affects the notion of an argument because a distinction needs to be made between what an argument is and the speech act of putting forward an argument for acceptance.

9. The Legal and the Everyday Notions of Presumption

For these reasons, it has clearly been shown that it is necessary to revise the philosophical accounts of presumption described in Chapter 1 (Section 4), and to see these accounts as presenting a definition of the speech act of putting forward a presumption in a dialogue. This revision leaves the question open of how to define the concept of presumption. The traditional approach portrays a presumption as something put forward in a dialogue, and offers a set of normative conditions defining how it should be put forward, and how the other party and the dialogue should properly react to its being put forward. However, this approach does not define what it is that has been put forward. The new theory dialogical of presumption outlined in Section 8 of Chapter 8 provides a different approach.

In presenting the dialogical theory, the following answer has been given to the question of how presumption is related to evidential burden. As explained earlier, the general principle of burden of proof requires that the party who makes a claim and puts forward an argument for its acceptance must supply evidence to back it up if the claim or argument is questioned. But it commonly happens that, for various reasons, it may be difficult or problematic to meet this requirement. It may be too costly to obtain such evidence, or even more generally, it may take so much time and effort to obtain it that this quest would obstruct the progress of the dialogue currently moving forward in its argumentation stage. In some instances, presumptive reasoning can be the most useful tool of choice in overcoming this problem. In such cases, raising a presumption can be a way to, if not meet the evidential burden, at least satisfy the need to meet it by justifying the drawing of a conclusion on a tentative basis subject to later retraction if contravening evidence comes to be known.

But now we come back to the criticism of hasty transference described in Chapter 8, Section 1. If the notion of presumption is as vague and shadowy, and as often exploited by judges as Gaskins and others claim, how can we justify the transference from the way presumptions work in the law to the way presumption is supposed to work in everyday conversational argumentation? There certainly is something in the criticism of hasty transference. There is plenty of evidence to back up the claim that the way that presumptions and burdens of proof are used as evidentiary devices by the courts constitutes what can properly be described as "a conceptual disarray" (Allen 1980, 323). Allen (1981, 845) commented that the term "presumption" has been used in such a widely varied way in judicial decision making that it has become merely "a label that has been applied to a widely disparate set of decisions concerning the proper mode of trial and the manner in which facts are to be established for the purpose of resolving legal disputes." The admittedly slippery and ambiguous usages of the devices of burden of proof and presumption in law supports the hasty transference criticism that trying to apply these legal notions to everyday conversational notions of burden of proof and presumption is not the right direction to take.

But here we need to recall Wigmore's view that underlying both evidential reasoning in law and evidential reasoning in everyday conversational contexts outside of law, there is what he called the science of evidence, meaning the logical structure of evidential reasoning. On Allen's view (1981, 845), the decisions arrived at by invoking the term presumption can also be reached by relying on "normal evidentiary concepts and policies." However, instead of dwelling on the ambiguity caused by the overly stretched use of this term, it is better to concentrate on the core evidential notion of presumption of the kind defined in rule 301 of the Federal Rules of Evidence quoted from the Cornell Web site.[1]

[1] http://www.law.cornell.edu/rules/fre/rule_301.

In a civil case, unless a federal statute or these rules provide otherwise, the party against whom a presumption is directed has the burden of producing evidence to rebut the presumption. But this rule does not shift the burden of persuasion, which remains on the party who had it originally. (Pub. L. 93–595, §1, Jan. 2, 1975, 88 Stat. 1931; Apr. 26, 2011, eff. Dec. 1, 2011)

Once the distinction between the burden of persuasion and the burden of producing evidence has been clarified at an appropriate level of logical abstraction using the formal systems of ASPIC+ and the Carneades Argumentation System, presumption becomes a clear and precise enough notion that it is useful for argumentation studies. This statement is not meant to represent the claim that the definition of "presumption" proposed in Chapter 3 represents all the subtleties and refinements in the various usages of the term that one can find in court rulings and other judicial sources. It is only meant to represent the claim that a core notion of presumption can be defined in terms of the account of burden of persuasion and burden of production of evidence set out in the theory of burden of proof in evidential reasoning of Chapter 2. Following this path, it has been argued that there can be a carefully modified transference from law that is not hasty, but that builds on the judicial wisdom of practices of common law. It can build on insights from the codification of evidential rules embodied in the Federal Rules of Evidence, and transfer it to an analysis of the notion of presumption that can be found in everyday conversational argumentation practices.

10. Conclusions and Suggestions for Further Research

It is a general conclusion of this book is that the notions of burden of proof and presumption work in a similar way in different settings of argumentation, for example, legal argumentation, forensic debate and everyday conversational argumentation of a largely unregulated sort. The approach of the book is meant to be flexible, in order to show how burden of proof operates in many different contexts of argumentation that are important to the concerns that the scholars in this field have shown. Thus, it has been contended that burden of proof operates in a different way in deliberation than it does it of persuasion dialogue or in an inquiry type of dialogue. Still, the central core of the model of burden of proof and presumption extracted from legal argumentation through the framework of AI, it has been argued, presents a central structure and a set of tools that can be usefully applied to all these different settings where it can be used to solve problems that arise from burden of proof.

These settings range from ones that are procedurally highly organized, for example legal argumentation in a trial setting, to those that are less organized, for example forensic debate, to those that seem to be least organized, for example everyday conversational arguments. Dealing with the latter kinds of arguments seems to be the greatest challenge because the protocols are not explicitly stated and agreed to at the opening stage. They have to be implicit maxims of politeness (Grice, 1975), or understood agreements that can form the basis of objections that can have force. For example, if a participant doesn't stick to the topic and tries to distract the audience by moving to a more emotionally exciting subject that is not relevant to the issue supposedly being discussed, other participants can accuse him of committing a fallacy of relevance. Indirectly, the force of this criticism implies that there is an implicit procedural agreement that if the parties are having a critical discussion, they should be bound to stick to the issue of the discussion and only use relevant arguments. Relevant arguments are taken to be those that carry probative weight in either proving or disproving the ultimate claim that is being disputed.

In the dialogue on tipping, there was no referee who could step in and enforce some good procedural rules of debate if the participants wandered away from the topic to be discussed or tried to make other moves that were not helpful to moving the quality of persuasion dialogue along. For example, one party might have tried to intimidate the other party by continually trying to make it appear that this party had failed to meet his or her burden of proof by using a fallacious argument from ignorance. Or one party might have tried to dominate the conversation by digressing at great length, thereby preventing the other party from speaking at all. Possibly the host could have intervened, if such underhanded argumentation tactics were used by one party to try to unfairly get the best of the other. As it happened, there was no need for this kind of intervention, because both Bob and Helen were probatively reasonable with regard to their contributions to the discussion. Each of them nicely took turns to put forward arguments to support their own claim or to attack the previous argument put forward by the other side. Generally the argumentation did follow the abstract argumentation pattern where each party at his or her move put forward an argument that aimed to defeat the argument put forward by the other party at his or her previous move. But as we observed in our commentary on the discussion, there was some backtracking, and there were quite a few instances where Bob or Helen put forward a new argument to support his or her claim instead of attacking the argument of the other side that was just put forward.

It could be said with some justification that this example is a toy example that is not all that realistic because in a real instance of a typical sequence of argumentation of this sort when we might find a real case, the turns taken at each move might comprise several arguments, and as well include all kinds of comments, interjections and questions that are not arguments. In this kind of case, there would have to be quite a bit more cleaning up of the text to identify the arguments and put them into a sequence where they could be analyzed using the tools developed in this book. That doesn't prove to be a serious problem, however, because it is the normal task of interpretation and analysis that is typical of any attempt to evaluate a sequence of argumentation of the kind found in natural language discourse. There is generally a careful procedure of interpretation and analysis that has to be undertaken before the point where any evaluation of the arguments in the text can be undertaken. This kind of work tends to be a little easier to carry out in analyzing argumentation in legal cases in a trial setting because extraneous comments, irrelevant arguments and other kinds of interjections are discouraged or even forbidden by the procedural rules enforced by the court. The tipping example has already been cleaned up to a great extent so that it can be used to illustrate how burden of proof shifts without introducing the numerous complications and digressions that would normally have to be dealt with, and that would interfere with the capability of the example to illustrate the features of the burden of proof that are important for us to learn about.

What has been revealed by the book is the negative conclusion that research on burden of proof and presumption is in its infancy as far as providing precise logical models suitable for use in argumentation studies is concerned. Much more work needs to be done to apply the model explained and developed in this book to longer and more complex examples using case study techniques. Much more work needs to be done in legal research on refining the model and applying it to legal cases to deal with important fundamental questions like the distinction between rebuttable and irrebuttable presumptions in law. Much more work needs to be done on analyzing fallacies, and other problematic argumentation moves that are associated with and arise from problems about managing burdens of proof and presumptions. Because these notions, while everybody acknowledges their importance, have proved to be elusive, ambiguous, slippery and hard to define in the past, and have been closely associated with many of the most significant logical fallacies, there has been little to draw on in the past. Now we have at least a foothold on the subject. The need is to press forward and apply it to the many practical domains of argumentation.

Bibliography

Alexy, R. (1989). *A Theory of Legal Argumentation.* Oxford: Oxford University Press.

Allen, R. J. (1980). Structuring Jury Decisionmaking in Criminal Cases: A Unified Constitutional Approach to Evidentiary Devices, *Harvard Law Review,* 94(2), 321–368.

Allen R. J. (1981). Presumptions in Civil Actions Reconsidered, *Iowa Law Review,* 66, 1981, 843–867.

Allen, R. J. (1993). How Presumptions Should be Allocated: Constitutional Adjudication, the Demands of Knowledge, and Epistemological Modesty, *Northwestern University Law Review,* 88(1), 436–456.

Allen, R. J. (1994). Burdens of Proof, Uncertainty and Ambiguity in Modern Legal Discourse, *Harvard Journal of Law and Public Policy,* 17(3), 1994, 627–646.

Allen, R. J. and Callen, C. R. (2003). Teaching Bloody Instructions: Civil Presumptions and the Lessons of Isomorphism, *Quinnipiac Law Review,* 21, 933–960.

Allen, R. J. and Pardo, M. S. (2007). The Problematic Value of Mathematical Models of Evidence, *Journal of Legal Studies,* 36, 107–140.

Anderson, T., Shum, D. and Twining, W. (2005). *Analysis of Evidence,* 2nd edition. New York: Cambridge University Press.

Aristotle (1984). Posterior Analytics, *The Complete Works of Aristotle,* vol. 1. ed. J. Barnes, Princeton: Princeton University Press.

Armstrong, K. and Possley, M. (1999). The Verdict: Dishonor, *Chicago Tribune,* January 10, 1999, 1,12.

Ashford, H. A. and Risinger, D. M. (1969). Presumptions, Assumptions and Due Process in Criminal Cases: A Theoretical Overview, *The Yale Law Journal,* 79, 165–208.

Atkinson, K. and Bench-Capon, T. J. M. (2007). Argumentation and Standards of Proof, *Proceedings of the Eleventh International Conference on AI and Law (ICAIL 2007),* ed. Radboud Winkels, New York: ACM Press, 107–116.

Atkinson, K., Bench-Capon, T. J. M., and McBurney, P. (2005). A Dialogue Game Protocol for Multi-Agent Argument over Proposals for Action, *Autonomous Agents and Multi-Agent Systems,* 11(2), 153–171.

Atkinson, K., Bench-Capon, T. J. M., and McBurney, P. (2006). Computational Representation of Practical Argument, *Synthese,* 152(2), 157–206.

Atkinson, K., Bench-Capon, T. and Walton, D. (2013). Distinctive Features of Persuasion and Deliberation Dialogues, *Argument and Computation*, 4(2), 105–127.

Bailenson, J. (2001). Contrast Ratio: Shifting Burden of Proof in Informal Arguments, *Discourse Processes*, 32(1), 29–41.

Bailenson, J. and L. J. Rips (1996). Informal Reasoning and Burden of Proof, *Applied Cognitive Psychology*, 10(7), 3–16.

Bailey, R. (1999). Precautionary Tale, http://www.reason.com/news/show/30977. html (last accessed December 24, 2008).

Barnes, J. (1990). *The Toils of Skepticism*. Cambridge: Cambridge University Press.

Barth, E. M. and Krabbe, E. C. W. (1982). *From Axiom to Dialogue: A Philosophical Study of Logics and Argumentation*. Berlin: Walter de Gruyter.

Bench-Capon, T. J. M. (1997). Argument in Artificial Intelligence and Law. *Artificial Intelligence and Law*, 5(4): 249–261.

Bench-Capon, T. J. M. (2003). Persuasion in Practical Argument Using Value-based Argumentation Frameworks. *Journal of Logic and Computation*, 13(3), 429–448.

Bench-Capon, T. J. M., Doutre, S., and Dunne, P. E. (2007). Audiences in Argumentation Frameworks. *Artificial Intelligence*, 171(1), 42–71.

Bench-Capon, T. J. M. and Sartor, G. (2003). A Model of Legal Reasoning with Cases Incorporating Theories and Values. *Artificial Intelligence* 150: 97–143.

Bench-Capon, T. J. M., Doutre, S. and Dunne, P. E. (2008). Asking the Right Question: Forcing Commitment in Examination Dialogues. *Computational Models of Argument: Proceedings of COMMA 2008*, eds. P. Besnard, S. Doutre and A. Hunter. Amsterdam: IOS Press, 49–60.

Bex, F. J. (2011). *Arguments, Stories and Criminal Evidence – A Formal Hybrid Theory*. Springer: Dordrecht.

Bex, F. J., van den Braak, S.W., van Oostendorp, H., Prakken, H., Verheij B. and Vreeswijk, G. (2007). Sense–making Software for Crime Investigation: How to Combine Stories and Arguments, *Law, Probability and Risk* 6, 145–168.

Bex, F. J., van Koppen, P. J., Prakken, H., and Verheij, B. (2010). A Hybrid Formal Theory of Arguments, Stories and Criminal Evidence, *Artificial Intelligence and Law*, 18(10), 123–152.

Bex, F. J. and Walton, D. (2012). Burdens and Standards of Proof for Inference to the Best Explanation: Three Case Studies, *Law, Probability and Risk*, 11(2–3), 113–133.

Bex, F., Prakken, H., Reed, C. and Walton, D. (2003).Towards a Formal Account of Reasoning about Evidence, Argument Schemes and Generalizations, *Artificial Intelligence and Law*, 11(2–3), 125–165.

Black, E. and Hunter, A. (2007). A Generative Inquiry Dialogue System. *Sixth International Joint Conference on Autonomous Agents and Multi-agent Systems*, Honolulu, Hawaii, 1010–1017.

Black, E. and Hunter, A. (2009). An Inquiry Dialogue System, *Autonomous Agent and Multi-Agent Systems* 19(2), 173–209.

Brandom, R. (1994). *Making it Explicit*. Cambridge, MA: Harvard University Press.

Brewka, G. (2001). Dynamic Argument Systems: A Formal Model of Argumentation Processes Based on Situation Calculus, *Journal of Logic and Computation* 11(2), 257–282.

Brewka, G. and Gordon, T. F. (2010). Carneades and Abstract Dialectical Frameworks: A Reconstruction. In P. Baroni, F. Cerutti, M. Giacomin, and G.R. Simari, eds., *Computational Models of Argument: Proceedings of COMMA 2010*. Amsterdam: IOS Press, 3–12.

Brown, D. G. (1955). The Nature of Inference, *The Philosophical Review*, 64(3), 351–369.

Caminada, M. (2004). *For the Sake of Argument: Explorations into Argument-Based Reasoning*, PhD thesis, Free University of Amsterdam: http://icr.uni.lu/~martinc/publications/thesis.pdf.

Caminada, M. (2008). On the Issue of Contraposition of Defeasible Rules. *Computational Models of Argument: Proceedings of COMMA 2008*. eds. P. Besnard, S. Doutre, and A. Hunter. Amsterdam: IOS Press, 109–115.

Clark, K. L. (1978). Negation as Failure, *Logic and Data Bases*, ed. H. Gallaire and J. Minker, New York: Plenum Press, 293–322.

Cohen, L. J. (1977). *The Probable and the Provable*. Oxford: Oxford University Press.

Cohen, L. J. (1979). On the Psychology of Prediction: Whose is the Fallacy? *Cognition*, 7(4), 385–407.

Cohen, L. J. (1980). Some Historical Remarks on the Baconian Notion of Probability, *Journal of the History of Ideas*, 41 (2), 219–231.

DeMorgan, A. (1926). *Formal Logic*. The Open Court Company: London. Reprint of the original Taylor and Walton edition of 1847.

Dohnal, M. (1992). Ignorance and Uncertainty in Reliability Reasoning. *Microelectronics and Reliability*, 32(8), 1157–70.

Doutre, S., McBurney, P., Wooldridge, M., and Barden, W. (2006). Information-Seeking Agent Dialogs with Permissions and Arguments, *Fifth International Joint Conference on Autonomous Agents and Multiagent Systems*. Technical Report: www.csc.liv.ac.uk/research/techreports/tr2005/ulcs-05-010.pdf.

Dube, F. (2002). Judge Criticizes 'Hired Gun' Experts. *National Post* (Canada), November 18, 2002, A1,A5.

Dung, P. (1995).On the Acceptability of Arguments and its Fundamental Role in Nonmonotonic Reasoning, Logic Programming and n-person Games. *Artificial Intelligence*, 77(2), 321–357.

Dunne, P. E., Doutre, S., and Bench-Capon, T. J. M. (2005). Discovering Inconsistency through Examination Dialogues. *Proceedings IJCAI-05, (International Joint Conferences on Artificial Intelligence*. Edinburgh, 1560–1561. Available at: http://ijcai.org/search.php.

Farley, A. M. and Freeman, K. (1995). Burden of Proof in Legal Argumentation', *5th International Conference on Artificial Intelligence and Law*, ed. T. J. M. Bench-Capon. New York: ACM Press, 156–164.

Feteris, E. (1999). *Fundamentals of Legal Argumentation*. Dordrecht: Kluwer.

Finocchiaro, M. (1980). *Galileo and the Art of Reasoning*. Dordrecht: Kluwer.

Finocchiaro, M. (2005). *Arguments about Arguments*. Cambridge: Cambridge University Press.

Fox, J. and Das, S. (2000). *Safe and Sound: Artificial Intelligence in Hazardous Applications*. Cambridge, MA: MIT Press.

Frank, J. (1963). *Courts on Trial*. New York: Atheneum.

Freeman, J. B. (1995). The Appeal to Popularity and Presumption by Common Knowledge, *Fallacies: Classical and Contemporary Readings*, eds. Hansen, H. V.

and Pinto, R. C., University Park: The Pennsylvania State University Press, 265–273.

Freeman, K. and Farley, A. M. (1996). A Model of Argumentation and Its Application to Legal Reasoning, *Artificial Intelligence and Law*, 4(3–4), 163–197.

Freestone, D. and Hey, E. (1996). Origins and Developments of the Precautionary Principle, *The Precautionary Principle and International Law*, eds. David Freestone and Ellen Hey. The Hague: Kluwer Law International, 3–15.

Garner, B. A. (1990). *Black's Law Dictionary*, 9th ed. St Paul, Minn: Thomson Reuters.

Gaskins, R. H. (1992). *Burdens of Proof in Modern Discourse.* New Haven: Yale University Press.

Godden, D. M. and Walton, D. (2007). A Theory of Presumption for Everyday Argumentation, *Pragmatics and Cognition*, 15(2), 2007, 313–346.

Gordon, T. F. (1995). *The Pleadings Game: An Artificial Intelligence Model of Procedural Justice.* Dordrecht: Kluwer Academic Publishers.

Gordon, T. F. (2008). Hybrid Reasoning with Argumentation Schemes, *CMNA 08: Workshop on Computational Models of Natural Argument.* Patras, Greece.

Gordon, T. F. (2010). An Overview of the Carneades Argumentation Support System, *Dialectics, Dialogue and Argumentation*, eds. C. Reed and C. W. Tindale, London: College Publications, 145–156.

Gordon, T. F. and Karacapilidis, N. (1997).The Zeno Argumentation Framework, *Proceedings of the Sixth International Conference on AI and Law*, ed. L. Karl Branting. New York: ACM Press, 10–18.

Gordon, T. F. and Walton, D. (2006). The Carneades Argumentation Framework, *Computational Models of Argument: Proceedings of COMMA 2006*, eds. P. E. Dunne and T. J. M. Bench-Capon. Amsterdam: IOS Press, 195–207.

Gordon, T. F. and Walton, D. (2006a). Pierson v. Post Revisited, *Computational Models of Argument: Proceedings of COMMA 2006*, eds. P. E. Dunne and T. J. M. Bench-Capon. Amsterdam: IOS Press, 208–219.

Gordon, T. F. Prakken, H., and Walton, D. (2007). The Carneades Model of Argument and Burden of Proof, *Artificial Intelligence*, 171(10–15), 875–896.

Gordon, T. F. and Walton, D. (2009). Proof Burdens and Standards. *Argumentation and Artificial Intelligence*, eds. I. Rahwan and G. Simari. Berlin: Springer, 239–260.

Governatori, G. (2008). Defeasible Logic: http://defeasible.org

Grice, H. P. (1975). Logic and Conversation, in D. Davidson and G. Harman, eds., *The Logic of Grammar.* Encino, CA: Dickenson, 64–75.

Hage, J. (2000). Dialectical Models in Artificial Intelligence and Law. *Artificial Intelligence and Law*, 8(2–3): 137–172.

Hage, J. C., Leenes, R., and Lodder, A. R. (1993). Hard Cases: A Procedural Approach *Artificial Intelligence and Law* 2(2): 113–167.

Hahn, U., and Oaksford, M. (2007). The Burden of Proof and its Role in Argumentation. *Argumentation*, 21(1), 39–61.

Hamblin, C. L. (1970). *Fallacies.* London: Methuen.

Hamblin, C. L. (1971). Mathematical Models of Dialogue. *Theoria* 37(2), 130–155.

Hansen, H. V. (2003). Theories of Presumptions and Burdens of Proof, in J. A. Blair, D. Farr, H. V. Hansen, R. H. Johnson, and C. W. Tindale, eds., *Informal Logic at 25: Proceedings of the Windsor Conference.* Windsor, Ontario: OSSA.

Hastings, A. C. (1963). *A Reformulation of the Modes of Reasoning in Argumentation.* Ph.D. diss., Northwestern University, Evanston, IL.

Hathcock, J. N. (2000). The Precautionary Principle – an Impossible Burden of Proof for New Products, *AgBioForum*, 3(4), 255–258.

Hoenig, M. (2002). Products Liability: Jury Rejection of Uncontradicted Expert Testimony. *New York Law Journal*, 228: 3–6.

James, F. (1961). Burdens of Proof, *Virginia Law Review*, 47(1), 1961, 51–70.

Kauffeld, F. J. (1998). Presumptions and Distribution of Argumentative Burdens in Acts of Proposing and Accusing. *Argumentation*, 12(2), 245–266.

Kauffeld, F. (2003). The Ordinary Practice of Presuming and Presumption with Special Attention to Veracity and Burden of Proof, *Anyone Who Has a View: Theoretical Contributions to the Study of Argumentation*, ed. van Eemeren, F. H., J. A. Blair, C. A. Willard, and A. F. Snoek Henkemans, Dordrecht: Kluwer, 136–146.

Keppens, J. (2009). Conceptions of Vagueness in Subjective Probability for Evidential Reasoning. In *Proceedings of the 22nd Annual Conference on Legal Knowledge and Information Systems*, ed. G. Governatori. Amsterdam: IOS Press, 89–99.

Kiralfy, A. (1987). *The Burden of Proof.* Abington: Professional Books.

Kienpointner, M. (1992). *Alltagslogik: Struktur und Funktion von Argumentationsmustern.* Stuttgart: Fromman-Holzboog.

Krabbe E. C. W. (1995). Appeal to Ignorance, in H. V. Hansen and R. C. Pinto, eds., *Fallacies: Classical and Contemporary Readings*, University Park: Pennsylvania State University Press, 251–264.

Krabbe, E. C. W. (1996). Can We Ever Pin One Down to a Formal Fallacy? In van Benthem, J., van Eemeren, F. H., and Frank Veltman, eds., Logic and Argumentation, Amsterdam: North-Holland, 129–141.

Krabbe E. C. W. (1999). Profiles of Dialogue, in J. Gerbrandy, Marx, M., de Rijke, M. and Venema, Y., eds., *JFAK: Essays Dedicated to Johan van Benthem on the Occasion of his 50th Birthday*, Amsterdam: Amsterdam University Press, 25–36.

Krabbe, E. C. W. (2001). The Problem of Retraction in Critical Discussion. *Synthese* 127(1–2), 141–159.

Krabbe, E. C. W. (2003). Metadialogues, *Anyone Who Has a View*, ed. Van Eemeren, F. H. et al., Dordrecht: Kluwer, 83–90.

Krabbe, E. C. W. (2007). On How to Get Beyond the Opening Stage, *Argumentation*, 21(3), 2007, 233–242.

Krabbe, E. C. W. (2013). Topical Roots of Formal Dialectic. *Argumentation*, 27(1), 71–87.

Lascher, E. L. (1999). *The Politics of Automobile Insurance Reform: Ideas, Institutions, and Public Policy in North America.* Washington: Georgetown University Press.

Laudan, L. (2006). *Truth, Error and Criminal Law.* Cambridge: Cambridge University Press.

Lodder, Arno R. (1999). *Dialaw: On Legal Justification and Dialogical Models of Argumentation.* Dordrecht: Kluwer.

Loftus, E. F. (2003). Our Changeable Memories: Legal and Practical Implications. *Neuroscience*, 4(3), 231–234.

Loftus, E. F. (2005). Planting Misinformation in the Human Mind: A 30-year Investigation of the Malleability of Memory. *Learning & Memory*, 12(4), 361–366.

Loui, R. P. (1998). Process and Policy: Resource-Bounded Nondemonstrative Reasoning. *Computational Intelligence* 14(1), 1–38.

Macagno, F. and Walton, D. (2011). Reasoning from Paradigms and Negative Evidence, *Pragmatics and Cognition*, 19(1), 92–116.

Mackenzie, J. D. (1979). How to Stop Talking to Tortoises, *Notre Dame Journal of Formal Logic* 20(4), 705–717.

Mackenzie, J. D. (1981). The Dialectics of Logic. *Logique et Analyse*, 24, 159–177.

Mackenzie, J. (1990). Four Dialogue Systems. *Studia Logica* 49(4), 567–583.

McBurney, P. and Parsons, S. (2001). Chance Discovery Using Dialectical Argumentation. *New Frontiers in Artificial Intelligence*, ed. T. Terano, T. Nishida, A. Namatame, S. Tsumoto, Y. Ohsawa, and T. Washio *Lecture Notes in Artificial Intelligence*, vol. 2253, Berlin: Springer Verlag, 414–424.

McBurney, P., Hitchcock, D. and Parsons, S. (2007). The Eightfold Way of Deliberation Dialogue, *International Journal of Intelligent Systems*, 22(1), 95–132.

Morman, D. (2005). The Wild and Wooly World of Inference and Presumptions – When Silence is Deafening, *The Florida Bar Journal*, 79(10), 38–43.

Nute, D. (2001). Defeasible Logic: Theory, Implementation, and Applications. *Proceedings of INAP 2001, 14th International Conference on Applications of Prolog*, IF Computer Japan, Tokyo, 87–114.

Nute, D. (1994). Defeasible Logic. *Handbook of Logic in Artificial Intelligence and Logic Programming, volume 3: Nonmonotonic Reasoning and Uncertain Reasoning*. Oxford: Oxford University Press, 353–395.

Pardo M. S. and Allen, R. J. (2007). Juridical Proof and the Best Explanation, *Law and Philosophy* 27(3), 223–268.

Park, R. C., Leonard, D. P., and Goldberg, S. H. (1998). *Evidence Law*, St. Paul: Minnesota: West Group.

Parsons, S. and Jennings N. (1997). Negotiation through Argumentation: A Preliminary Report. *Proceedings of the Second International Conference on Multi-Agent Systems*, ed. Mario Tokoro. Menlo Park, CA: AAAI Press, 267–274.

Peirce, C. S. (1934a). *Belief and Judgement*. Cambridge: Harvard University Press.

Peirce, C. S. (1934b). *Collected Papers*, vol V. Cambridge: Harvard University Press.

Perelman, C. and Olbrechts-Tyteca, L. (1971). The New Rhetoric: A Treatise on Argumentation. In J. Wilkinson and P. Weaver (trans.), 2nd edition, Notre Dame, IN: University of Notre Dame Press. (First published, as La Nouvelle Rhetorique, in 1958.)

Pollock, J. L. (1995). *Cognitive Carpentry*. Cambridge: MA, The MIT Press.

Prakken, H. (1997). *Logical Tools for Modelling Legal Argument*. Dordrecht: Kluwer.

Prakken, H. (2001). Modelling Defeasibility in Law: Logic or Procedure, *Fundamenta Informaticae* 48(2), 253–271.

Prakken, H. (2001a). Relating Protocols for Dynamic Dispute with Logics for Defeasible Argumentation. *Synthese* 127(1–2), 187–219.

Prakken, H. (2002). Models of Dispute Resolution: A Formal Framework and an Application. www.cs.uu.nl/staff/henry.html, 1–69.

Prakken, H. (2005). Coherence and Flexibility in Dialogue Games for Argumentation. *Journal of Logic and Computation* 15(6), 1009–1040.

Prakken, H. (2006). Formal Systems for Persuasion Dialogue. *The Knowledge Engineering Review*, 21(2), 163–188.

Prakken, H. and G. Sartor (1997). Argument-based Extended Logic Programming with Defeasible Priorities, *Journal of Applied Non-classical Logics,* 7(1–2), 25–75.

Prakken, H. and Sartor, G. (2006). Presumptions and Burdens of Proof, *Legal Knowledge and Information Systems: JURIX 2006: The Nineteenth Annual Conference,* ed. T. M. van Engers, Amsterdam: IOS Press, 21–30.

Prakken, H. and Sartor, G. (2007). Formalising Arguments about the Burden of Persuasion. *Proceedings of the Eleventh International Conference on Artificial Intelligence and Law.* New York: ACM Press, 97–106.

Prakken, H. and Sartor, G. (2009). A Logical Analysis of Burdens of Proof. *Legal Evidence and Burden of Proof,* ed. Kaptein, H., Prakken, H., and Verheij, B. Farnham: Ashgate, 223–253.

Prakken, H., Reed, C., and Walton, D. (2005). Dialogues about the Burden of Proof, Proceedings of the Tenth International Conference on AI and Law, June 6–11, Bologna, Italy. New York: The Association for Computing Machinery, 115–124.

Prakken, H. and Sartor, G. (2011). On Modelling Burdens and Standards of Proof in Structured Argumentation. In Atkinson, K. D. (ed.), *Legal Knowledge and Information Systems. JURIX 2011: The Twenty-Fourth Annual Conference.* Amsterdam: IOS Press, 83–92.

Recuerda, M. (2008). Dangerous Interpretations of the Precautionary Principle and the Foundational Values of the European Food Law: Risk versus Risk, *Journal of Food Law and Policy,* 4(1), 1–44.

Reed, C. (2006). Representing Dialogic Argumentation. *Knowledge-Based Systems,* 19, 22–31.

Reed, C. and Rowe, G. (2004) Araucaria: Software for Argument Analysis, Diagramming and Representation, *International Journal on Artificial Intelligence Tools,* 13(4), 961–979.

Reed, C. and Walton, D. (2003). Applications of Argumentation Schemes. In *Argumentation and Its Applications: Proceedings of the 4th OSSA Conference,* ed. Hansen, H. V., Tindale, C. W., Blair, J. A., and Johnson, R. H., Windsor, Canada.

Reed, C. and Walton, D. (2003a). Diagramming, Argumentation Schemes and Critical Questions. *Proceedings of the Fifth International Conference of the International Society for the Study of Argumentation,* ed., van Eemeren, F. H., et al. Amsterdam: Sicsat, 881–886.

Reed, C. and Walton, D. (2003a). Argumentation Schemes in Argument-as-Process and Argument-as-Product, *Proceedings of the Conference Celebrating Informal Logic @25,* eds., Blair, J. A., Farr, D., Hansen, H. V., Johnson, R. H. and Tindale, C. W. Ontario Society for the Study of Argumentation. Windsor, Ontario. (CD-ROM.)

Reed, C. and Walton, D. (2007). *Argumentation Schemes in Dialogue, Dissensus and the Search for Common Ground: Proceedings of OSSA.* ed. Hansen, H. V., Tindale, C. W., Blair, J. A., Johnson, R. H., and Godden, D. M. Windsor, Ontario (CD-ROM.)

Reiter, R. (1980). A Logic for Default Reasoning, *Artificial Intelligence,* 13(1–2), 81–132.

Reiter, R. (1987). Nonmonotonic Reasoning, *Annual Review of Computer Science,* 2, 147–186.

Rescher, N. (1964). *Introduction to Logic.* New York: St. Martin's Press.

Rescher, N. (1976). *Plausible Reasoning.* Assen: Van Gorcum.

Rescher, N. (1977). *Dialectics.* Albany: State University of New York Press.

Rescher, N. (2006). *Presumption and the Practices of Tentative Cognition.* Cambridge: Cambridge University Press.

Rescorla, M. (2009a). Shifting the Burden of Proof? *The Philosophical Quarterly,* 59, 86–109.

Rescorla, M. (2009b). Assertion and its Constitutive Norms. *Philosophy and Phenomenological Research,* 79(1), 98–130.

Ricco, Robert, B. (2011). Individual Differences in Distinguishing Licit from Illicit Ways of Discharging the Burden of Proof, *Journal of Pragmatics,* 43(2), 616–631.

Safer, Jay G. (2002). Evolving Standards on Expert Witnesses. *New York Law Journal* 228: s4–s12.

Sartor, G. (1993). Defeasibility in Legal Reasoning. *Rechtstheorie,* 24, 281–316.

Schaffer v. Weast, Supreme Court of the United States, 2005. U.S. Lexis 8554, decided November 14, 2005.

Schank, R. C. (1986). *Explanation Patterns: Understanding Mechanically and Creatively.* Hillsdale, New Jersey: Erlbaum, 1986.

Schank, R. C. and Abelson, R. P. (1977). *Scripts, Plans, Goals and Understanding.* Hillsdale, NJ: Erlbaum.

Schank, R. C. and Riesback, C. K. (1981). *Inside Computer Understanding* Hillsdale, NJ: Erlbaum.

Schank, R. C., Kass, A., and Riesbeck, C. K. (1994). *Inside Case-Based Explanation.* Hillsdale, NJ: Erlbaum.

Schum, D. A. (1994).*Evidential Foundations of Probabilistic Reasoning,* New York: John Wiley and Sons.

Searle, J. R. (2001). *Rationality in Action.* Cambridge: The MIT Press.

Sidgwick, A. (1883). *Fallacies.* London: Kegan Paul. (Pages cited are from a character recognition reprint by General Books, 2009.)

Sidgwick, A. (1884). *Fallacies.* New York: D. Appleton and Company.

Singh, P., Lin, T., Mueller, E. T. Lim, G., Perkins, T. and Zhu, W. L. (2002). Open Mind Common Sense: Knowledge Acquisition form the General Public. *Proceedings of the First International Conference on Ontologies, Databases, and Applications of Semantics for Large Scale Information Systems,* Lecture Notes in Computer Science, Heidelberg: Springer Verlag.

Smets, P. (1991). Varieties of Ignorance and the Need for Well-Founded Theories. *Information Sciences,* 57–58, 135–144.

Stein, A. (2005). *Foundations of Evidence Law.* Oxford: Oxford University Press.

Strong, J. W. (1992). *McCormick on Evidence,* 4th edition, St Paul, MN: West Publishing Co. [Charles Tilford McCormick (1889–1963) was the original author, but this book has been rewritten by Strong and a team of six co-authors. It is referred to as (Strong, 1992).]

Tang, Y. and Parsons, S. (2006). Argumentation-Based Multi-agent Dialogues for Deliberation, *Proceedings of ArgMAS 2005, Lecture Notes in Artificial Intelligence 4049.* Berlin: Springer, 229–244.

Thayer, J. B. (1898). *A Preliminary Treatise on Evidence at the Common Law.* Boston: Little, Brown, and Company.

Tillers, P. (1989). Webs of Things in the Mind: A New Science of Evidence, *Michigan Law Review,* 87(6), 1225–1258.

Tillers, P. and Gottfried, J. (2006). United States v. Copeland: A Collateral Attack on the Legal Maxim that Proof Beyond a Reasonable Doubt Is Unquantifiable?, *Law, Probability and Risk*, 5, 135–157.

Tindale, C. (2010). Reason's Dark Champions: Constructive Strategies of Sophistic Argument. Columbia: The University of South Carolina Press.

Toulmin, S. (1964). *The Uses of Argument.* Cambridge: Cambridge University Press.

Twining, W. (1985). *Theories of Evidence: Bentham and Wigmore.* London: Weidenfeld and Nicolson.

Ullman-Margalit, E. (1983). On Presumption, *Journal of Philosophy*, 80(3), 143–163.

Van Eemeren, F. H. and Grootendorst, R. (1983). *Speech Acts in Argumentative Discussions.* Dordrecht: Foris Publications.

Van Eemeren, F. H. and Grootendorst, R. (1987). Fallacies in Pragma-Dialectical Perspective, *Argumentation*, 1(3), 283–301.

Van Eemeren, F. H. and Grootendorst, R. (1992). *Argumentation, Communication and Fallacies.* Mahwah, NJ.: Erlbaum.

Van Eemeren F. H. and Grootendorst, R. (2004), *A Systematic Theory of Argumentation: The Pragma-dialectical Approach.* Cambridge: Cambridge University Press.

Van Eemeren, F. H. and Houtlosser, P. (2002). Strategic Maneuvering with the Burden of Proof, *Advances in Pragma-Dialectics*, ed. van Eemeren, F. H. Amsterdam: SicSat, 2002, 13–28.

Van Gelder, T. J. (2007). The Rationale for Rationale™. *Law, Probability and Risk*, 6(1–4), 23–42

Van Gijzel, B. and Prakken, H. (2011). Relating Carneades with Abstract Argumentation. *Proceedings of the 22nd International Joint Conference on Artificial Intelligence (IJCAI 2011)*, 1113–1119. Barcelona,Spain.

van Laar, J. A. (2003). *The Dialectic of Ambiguity: A Contribution to the Study of Argumentation*, Ph.D. dissertation, Groningen University, Groningen, the Netherlands.

van Laar, J. A. and Krabbe, E. C. W. (2013). The Burden of Criticism: Consequences of Taking a Critical Stance. *Argumentation*, 27, 201–224.

Verheij, B. (1996). Rules, Reasons, Arguments: Formal Studies of Argumentation and Defeat. PhD diss., University of Maastricht, the Netherlands.

Verheij, B. (1999). Logic, Context and Valid Inference, *Legal Knowledge Based Systems. JURIX 1999*, eds. van den Herik, H. J. et al. Nijmegen: Gerard Noodt Instituut, 109–121.

Verheij, B. (2003). Dialectical Argumentation with Argumentation Schemes: An Approach to Legal Logic, *Artificial Intelligence and Law*, 11(2–3), 167–195.

Verheij, B. (2005). *Virtual Arguments. On the Design of Argument Assistants for Lawyers and Other Arguers.* The Hague: TMC Asser Press.

Walton, D. (1984). *Logical Dialogue-Games and Fallacies.* Lanham, MD: University Press of America.

Walton, D. (1988). Burden of Proof, *Argumentation*, 2(2), 233–254.

Walton, D. (1989). *Informal Logic.* Cambridge: Cambridge University Press.

Walton, D. (1992). *Plausible Argument in Everyday Conversation.* Albany: State University of New York Press.

Walton, D. (1992a). Commitment, Types of Dialogue and Fallacies, *Informal Logic*, 14, 1993, 93–103.

Walton, Douglas N. (1995). *A Pragmatic Theory of Fallacy.* Tuscaloosa: University of Alabama Press.

Walton, D. (1996). *Arguments from Ignorance.* University Park: Penn State Press.

Walton, D. (1996a). *Argumentation Schemes for Presumptive Reasoning.* Mahwah, NJ: Erlbaum.

Walton, D. (1997). *Appeal to Expert Opinion.* University Park: Pennsylvania State Press.

Walton, D. (1998). *The New Dialectic: Conversational Contexts of Argument.* Toronto: University of Toronto Press.

Walton, D. (2002). Are Some Modus Ponens Arguments Deductively Invalid?, *Informal Logic,* 22(1), 19–46.

Walton, D. (2002a). *Legal Argumentation and Evidence.* University Park: Pennsylvania State University Press.

Walton, D. (2003). Is there a Burden of Questioning?, *Artificial Intelligence and Law,* 11(1), 2003, 1–43.

Walton, D. (2006). How to Make and Defend a Proposal in Deliberation Dialogue. *Artificial Intelligence and Law,* 14(3), 177–239.

Walton, D. (2006a). Argument from Appearance: A New Argumentation Scheme, *Logique et Analyse,* 49(195), 319–340.

Walton, D. (2006b). Examination Dialogue: An Argumentation Framework for Critically Questioning an Expert Opinion, *Journal of Pragmatics,* 38(5), 745–777.

Walton, D. (2006c). *Fundamentals of Critical Argumentation.* Cambridge: Cambridge University Press.

Walton, D. (2007). *Dialog Theory for Critical Argumentation.* Amsterdam: John Benjamins Publishers.

Walton, D. (2007a). Metadialogues for Resolving Burden of Proof Disputes, *Argumentation,* 21(3), 291–316.

Walton, D. (2008). *Witness Testimony Evidence.* Cambridge: Cambridge University Press.

Walton, D. (2010). Burden of Proof in Deliberation Dialogs, *Proceedings of ArgMAS 2009,* ed. P. McBurney et al., Lecture Notes in Computer Science 6057, Heidelberg, Springer, 1–22.

Walton, D. (2011). A Dialogue System Specification for Explanation, *Synthese,* 182(3), 349–374.

Walton, D., Atkinson, K., Bench-Capon, T. J. M., Wyner, A., and Cartwright, D. (2010). Argumentation in the Framework of Deliberation Dialogue, *Arguing Global Governance,* ed. Bjola, C. and Kornprobst, M. London: Routledge, 210–230.

Walton, D. and Godden, D. M. (2005). The Nature and Status of Critical Questions in Argumentation Schemes, *The Uses of Argument: Proceedings of a Conference at McMaster University, 18–21 May 2005,* ed. David Hitchcock, Hamilton, Ontario, Ontario Society for the Study of Argumentation, 476–484.

Walton D. and Gordon, T. F. (2005). Critical Questions in Computational Models of Legal Argument, *Argumentation in Artificial Intelligence and Law,* IAAIL Workshop Series, ed. Dunne, P. E. and Bench-Capon, T. Nijmegen: Wolf Legal Publishers,103–111.

Walton, D. and Krabbe, E. C. W. (1995). *Commitment in Dialogue.* Albany: State University of New York Press.

Walton, D. and Macagno, F. (2006). Common Knowledge in Argumentation, *Studies in Communication Sciences*, 6 (1), 3–26.

Walton, D., Reed, C. and Macagno, F. (2008). *Argumentation Schemes*. Cambridge: Cambridge University Press.

Walton, D. and Sartor, G. (2013) Teleological Justification of Argumentation Schemes, *Argumentation* 27(2), 111–142.

Whately, R. (1846). *Elements of Rhetoric*, 7th edition, ed. D. Ehninger. Carbondale, IL: Southern Illinois University Press. (Reprinted 1963.)

Whinery, L. H. (2001). Presumptions and their Effect, *Oklahoma Law Review*, 54(3), 553–571.

Wigmore, J. H. (1931). *The Principles of Judicial Proof.* 2nd edition, Boston: Little, Brown and Company.

Wigmore, J. H. (1935). *A Student's Textbook of the Law of Evidence.* Chicago: The Foundation Press.

Wigmore, J. H. (1940). *A Treatise on the Anglo-American System of Evidence in Trials at Common Law*, 3rd edition, vol. 1. Boston: Little, Brown and Company.

Wigmore, J. H. (1981). *Evidence in Trials at Common Law*, vol. 9, revised by J. H. Chadbourn. Boston: Little, Brown and Company.

Williams, C. R. (2003). Burdens and Standards in Civil Litigation. *Sydney Law Review*, 9(2), 165–188.

Williams, G. (1977). The Evidential Burden: Some Common Misapprehensions, *New Law Journal*, 153, 156–158.

Witte, C. L., Kerwin, A., and Witte, M. H. (1991). On the Importance of Ignorance in Medical Practice and Education. *Interdisciplinary Science Reviews* 16(4): 295–298.

Woods, J., Irvine, A., and Walton, D. (2000). *Argument: Critical Thinking, Logic and the Fallacies.* Toronto: Prentice Hall.

Wooldridge, M. J., McBurney, P., and Parsons, S. (2005): The Meta-Logic of Arguments. In Dignum, F., Dignum, V., Koenig, S., Kraus, S., Singh, M. P., and Wooldridge, M., eds., *Proceedings of the Fourth International Joint Conference on Autonomous Agents and Multi-Agent Systems (AAMAS 2005)*, Utrecht, The Netherlands. New York, NY, USA: ACM Press, 560–567.

Index